Preface

I commenced these drawings in the summer of 1984/85 attempting to depict some of the Shakespearean characters after whom the moons of Uranus were named. I planned then to incorporate Chinese elements in these drawings as I had become interested in Chinese culture 20 years earlier after returning to Australia from Europe and working in the Advanced Studies Library of the Australian National University. There I came in contact with many Asian people who worked in or frequented the university's Oriental Library and I developed an interest in Chinese civilization. In 1969 I enrolled in the Asian Studies Faculty of the National University hoping to learn Asian languages and teach them in Australian schools. I first studied Indonesian and Japanese as Chinese was then rarely taught in Australian schools. Finally in 1973 I began the study of Classical Chinese, and started to learn the Sān Zì Jīng 三字经 and read parts of the Lùn Yǔ 论语 of Confucius. I was also introduced to such traditional things as the 28 Lunar Mansions, the 24 Solar Terms, the 8 Trigrams and 64 Hexagrams as well as the 10 Heavenly Stems, the 12 Earthly Branches and the 60 Year Cycle. I began to think then of doing drawings of these and integrating them with the 88 constellations, the asteroids, planets and moons in a pack of playing cards.

It was in the 1980s while working in the University of Sydney Library that I had access to useful information on astronomy, Greco-Roman history and art, a prerequisite for someone wanting to depict the mythical heroes after whom the constellations, the planets and their moons and the asteroids were mostly named. At first I wrote about each of these hundreds of mythical or historical figures and then began drawing the 88 Constellations. In 2006 I drew the remainder of the moons of Uranus which by then had grown to 27 and the moons of Neptune followed. It was only in 2010 after retiring and living in China that I completed drawings of the moons of Jupiter, Saturn and the other planets followed by those Chinese cultural items. Thus it has taken 30 years to complete this project which now incorporates a Chinese version of Tarot.

John Oxenham Goodman, Shenyang City, Liaoning Province, China

A Brief Biography of the Author

I was born in Australia in 1941. As a young man I worked in clerical positions in government departments but found these jobs routine and unchallenging. I studied Spanish and German and then travelled extensively in Western Europe crossing into Soviet occupied Berlin and climbing a mountain in Norway to view the midnight sun. These and other travel experiences opened my eyes to the wider world and gave me a broader perspective than that available in the geographical remoteness of Australia.

On returning to Australia I found employment in the Australian National University Library. I developed an interest in Asian civilizations and in 1969 enrolled in the Asian Studies Faculty. I first studied Indonesian and Japanese and eventually, in early 1973 undertook an intensive course in spoken Chinese at the University of Canberra and then studied Classical Chinese at the Australian National University. I finished my Indonesian and Asian Studies majors and studied Javanese and Arabic while completing a reading course in Dutch. I graduated with a Bachelor of Arts Honours degree and later received graduate diplomas in Education and Librarianship going on to major in Japanese language at the University of New South Wales. Much later I studied the Teaching of English as a Second Language at the Australian Catholic University in Sydney.

I retired and started living in China in 2010 and found displays in China's provincial museums to be a wonderful source of information on the ancient past. I travelled by train to most of China's provinces paying special attention to Buddhist, Daoist and Confucian temples. There I saw statues, engravings, paintings and art work which, when added to the enlightenment gained from museum visits, gave me fresh understanding of this 5000 year old civilization.

John Oxenham Goodman, Shenyang City, Liaoning Province, China.

Astronomical Pencil Drawings with Chinese and Western Elements, Planets, Moons and Stars

天文学方面的铅笔画—包含中西方五行，行星，卫星及星辰之间的关联

This collection of over 1,000 pencil drawings by John Oxenham Goodman includes images of the gods and historical persons after whom the modern constellations, planets, moons and chemical elements were named. It also contains pencil sketches of ancient Chinese heavenly bodies such as the 10 Shang Dynasty Suns, the Seven Luminaries, the 28 Lunar Mansions, the 9 Northern Dipper Stars and the 6 Southern Dipper Stars thus giving Eastern and Western perspectives of astronomy. Finally at the end of the book the Five Elements of ancient China are employed as suits in a Chinese version of Tarot using Mahjong symbols and concepts from China's 13th Century Mǎ Diào playing cards. The seasons, directions, days of the week, 24 Solar Terms, yearly cycles, signs of the zodiac as well as the 48 Japanese Flower Cards, representing the months of the year, are also interrelated.

Volume 5 of a 5-volume series illustrating the mythology of the heavenly bodies and the calendar in ancient China and the West including a retelling of ancient legends and an analysis of historical events.

by

John Oxenham Goodman 约翰•奥克森那姆•古德曼

Sydney, 2016

Copyright 2015 © John Oxenham Goodman, Sydney, Australia
All rights reserved. No part of this publication may be reproduced, stored in a retrieval system, or transmitted in any form or by any means, electronic, mechanical, photocopying, recording or otherwise, without the prior written permission of the copyright owner John Oxenham Goodman.

First Revised Edition with corrections

The Heavens and the Passing of Time in Art, Myth and History — A 5-volume series illustrating the mythology of the heavenly bodies and the calendar in ancient China and the West. Relationships to modern astronomy and traditional Chinese views of the heavens as well as to the five elements and the eight trigrams are also included and the pencil drawings are accompanied by a retelling of ancient legends and an analysis of historical events.

Books by John Oxenham Goodman 约翰•奥克森那姆•古德曼

Volume 1—The Chemical Elements and the 88 Constellations in Art, Myth and History 历史神话艺术中的化学元素和88星座。Amazon, 2015

Volume 2—The Sun, Planets, Dwarf Planets and Moons of the Solar System in Art, Myth and History 历史神话艺术中的太阳系的太阳，行星，矮行星和卫星。Amazon, 2015.

Volume 3—A Selection of mainly Larger Asteroids, their Moons and Four Comets in Art, Myth and History 历史神话艺术中的大的小行星及其卫星和四个彗星。Amazon, 2015.

Volume 4—Tarot-Mahjong related to the 5 Elements, the 8 Trigrams and other traditional series 塔罗麻将牌与五行，八卦及其它传统元素的关联。Amazon, 2015.

Volume 5— Astronomical Pencil Drawings with Chinese and Western Elements, Planets, Moons and Stars 天文学方面的铅笔画—包含中西方五行，行星，卫星及星辰之间的关联。Amazon, 2016.

Pre-Columbian Discoveries of the New World by Asians, Africans and Europeans and some Ancient Native American Voyages to Europe, Africa and Asia 亚非欧洲人在哥伦布之前发现美洲新大陆，以及古代美国原住民的亚非欧航行之旅。 Amazon, 2015.

Chinese Nationalism and Politics in Indonesia 1900-1965 : 1900-1965 在印度尼西亚的华侨民族主义和政治运动。Amazon, 2016.

Christianity—a Restatement of Greco-Roman Beliefs 基督教—古希腊罗马信仰的另一种形式。Amazon, 2016

The Rise of the Mauryan Empire and India's Relations with the Ancient Greek World 孔雀王国的崛起以及印度和古希腊的关系。Amazon, 2016

Was Noah a Woman? 诺亚是不是女人？Amazon, 2016.

Revised Edition Note

This revision of a preliminary edition became necessary when changes were made to pages 263-285 connected with the origins of the mahjong tiles Zhōng Fā Bái 中发白 and their relationship with Zhōng Huā 中花, Lǎo Qiān 老千 and Bái Huā 白花 in the 18th Century Pèng Hé cards 碰和牌. The abandonment of the suit of Tens from 18th Century Mǎ Diào 马吊 cards is also a related matter discussed in Volume 4 of this series. A drawing of a female Chinese Venus from ancient times was added to page 22 and drawings of the ancient female military commander Fù Hǎo 妇好 and the ancient Maritime Silk Road replaced other drawings on pages 265 and 284. Images of potential dwarf planets Salacia and Varda and their moons were also added to pages 220-223.

Foreword

This book provides images of the gods and legendary heroes who gave their names to the constellations, planets, minor planets, moons and chemical elements and these images are drawn both from modern Western and ancient Chinese perspectives. Astronomers, who took months or even years to decide on names for heavenly bodies they discovered, might surely be interested in seeing these pencil drawings which number over 1000. The stars and planets are not only of interest to astronomers but to the whole of the human population of out earth. It should not be surprising then that a person who is not an astronomy graduate would take on the task of drawing these images. Astronomy can be popularized through art and the fascinating stories of gods, goddesses, legendary heroes and heroines.

This book is also a visual representation of the measurement of time using Chinese, Japanese and Western traditions. The 5 Elements of ancient China combine with the sun and moon to give us the names of the 7 days of the week which are still used in Japan. The Old English (Anglo-Saxon) and Roman gods are depicted as representing the days of the week in the Western calendar. Then the animals of the 28 Chinese Lunar Mansions complete the lunar month while the 12 months of the year are represented by the monthly flowers on the Japanese Flower Cards (Hanafuda). The 12 Earthly Branches represented by 12 animals give us a 12 year period and when combined with the 10 Heavenly Stems they give us the 60 year cycle of the traditional Chinese calendar. Then there are the 24 Solar Terms, the 12 signs of the zodiac, the 5 Seasons, Directions, Emperors and Mountains, the 10 Suns of the Shang Dynasty, the 9 Northern Dipper Stars, the 6 Southern Dipper Stars, the 8 Trigrams, and 64 Hexagrams, all related to a newly created Chinese version of Tarot-Mahjong.

Volumes 1-4 of this series relate the gods, mythical and historical figures, animals, and other objects in the drawings of Volume 5 to one another and explain the history and mythology surrounding them. The drawings have been produced in a format suitable for imprinting on playing cards.

Contents

Books by John Oxenham Goodman	ii
Revised Edition Note	iii
Foreword	iv
Preface	v
Brief Biography of the Author	vi
Arrangement and Numbering of the Drawings	ix
Symbols representing each Series of Drawings	xii
Relationship of the 5 Chinese Elements to Tarot Suits	xvi
The 5 Chinese Elements	1
The 4 Ancient Greek Elements and Aether (or Spirit)	3
The 4 Seasons and the Change of Seasons	5
The 5 Emperors related to the 5 Planets	7
The 5 Mountains of Daoism and their Diagrams	9
The Animals of the 5 Directions	11
The English Days of the Week	13
The Roman Days of the Week	15
The Chinese and Japanese Days of the Week	18
The 7 Chinese Luminaries	20
The 28 Chinese Lunar Mansions	24
The 24 Solar Terms	32
The 12 Months of the Japanese Flower Cards	39
The 12 Signs of the Western Zodiac	51
The 10 Heavenly Stems	55
The 12 Earthly Branches	58
The 60 Year Cycle	61
Wújí, Tàijí, the 2 Forms and the 4 Phenomena	77
The Earlier Heaven Trigrams (Bā Guà)	80
The Later Heaven Trigrams (Bā Guà)	82
The Martial Arts Trigrams (Bā Guà)	85
The 64 Hexagrams	87
The 9 Northern Dipper Stars	104
The 6 Southern Dipper Stars	107
The 9 Sons of the Dragon	109
The Chemical Elements	112

The 88 Constellations	141

The Sun and Planets
The 10 Shang Dynasty Suns	164
The Sun, Planets and Moons to Ceres	167
Jupiter and its Moons	170
Saturn and its Moons	188
Uranus and its Moons	204
Neptune and its Moons	213
Dwarf Planets and Potential Dwarf Planets Orcus to Sedna	217
Asteroids and Moons, Comets	225

Tarot-Mahjong
Suit 1, Wood (Air), Swords/Spades	263
Suit 2, Fire, Batons/Clubs	267
Suit 3, Earth, Coins/Diamond	271
Suit 4, Metal (Aether/Spirit), Major Arcana/Towers	275
Suit 5, Water, Cups/Hearts, *Wan*	281
The 4 Great Heavenly Kings and the Chinese Earth God	285
The 5 Great Ancient Capitals of China	287
The 4 Seasons	289
The 4 Flowers	291
The 4 Pastimes	292
The 4 Professions	293
The 4 Hong Kong Mahjong Animals	294
The 5 Stars of Good Fortune	295
The 5 Animals of the Fortune Stars	297
The 5 Symbols of Prosperity	299
The 5 Auspicious Animals	301
The 8 Immortals	303
The 8 Ritual Implements of the Immortals	305

Chinese Chess
Chinese Chess	308

Index
	312

Arrangement and Numbering of the Drawings

The drawings in this 5-volume series illustrating the mythology of the heavenly bodies, the Eastern and Western calendar, the 4 Greek and 5 Chinese Elements and related concepts have been done in a size suitable for printing on playing cards. The cards are shaped like the diagram below with numbers placed along the side columns and top margin and with the drawing in the central space. For example, each drawing of a constellation is numbered in C from 1 to 88 while the drawing is placed

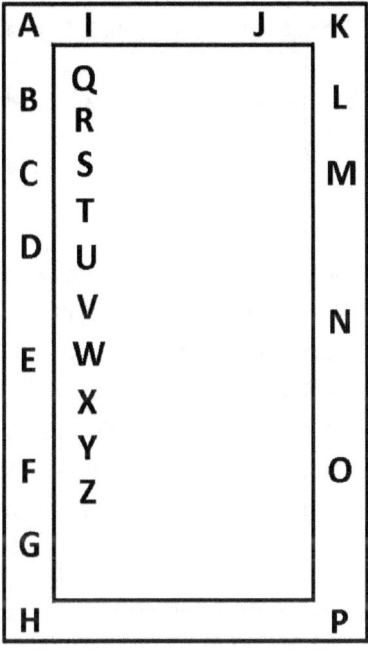

in the large space in the middle of the card. Similarly, each drawing of a planet is numbered sequentially in E while the drawing is in the central space. Additionally a series title such as "Constellations" or "Planets and Moons" is placed above the top margin and the card's number is repeated there. This system is continued from A through to P while additional series (Q to Z) are numbered and illustrated above the top margin and/or in the central space. A to I are largely based on Western science, astronomy and mythology while J to Z are mostly based on the Chinese and Japanese calendar, astronomy and mythology. In short, the left side of the diagram is largely Western while the right side is mostly Eastern although there is some mixing of the two.

A The 112 modern Chemical Elements [Volume 1]
B The directions North, East, South, West and Centre plus NE, SE, SW and NW [Volume 4]
C The 88 modern Constellations [Volume 1]
D The 10 Suns. This is a series of 10 cards, representing the 10 Shang Dynasty suns, 9 of which were shot down by the mythical Hòu Yì 后羿. The 10 suns are related to the 10 Heavenly Stems. [Volume 4]
E The Greco-Roman sun god Helios, 8 Planets, 14 Dwarf Planets and Potential Dwarf Planets and the 184 moons orbiting them (Earth 1, Mars 2, Jupiter 67, Saturn 62, Uranus 27, Neptune 13, Orcus 1, Pluto 5, Salacia 1, Haumea 2, Quaoar 1, Varda 1 and Eris 1) [Volume 2]
F 120 Asteroids (mostly with diameters of 100 kilometres or more) and their moons plus 4 Comets [Volume 3]
G Cards from among the Tarot/Mahjong series which are related to other cards [Volume 4]
H The 12 Signs of the Western Zodiac [Volume 4]
I Cards 1 to 14 of each of the four Tarot suits (Coins, Swords, Sticks or Batons and Cups) and Cards 0 to 22 of the Major Arcana are named here in English and Chinese. Three of the Tarot suits (Coins, Sticks and Cups) are identified with the 3 Mahjong suits (Coins 筒(饼), Strings of Coins 索(條) and *Wan* or 10,000 coins 万(萬)). However, the corresponding playing card images of Diamonds, Spades, Clubs and Hearts, which are easy to recognize, appear on the top left hand side of each of the four Tarot suits and an image of a tower appears at the top left of each Major Arcana card. [Volume 4]
J The 5 Chinese Elements (or Agents), the 5 Directional Animals, the 4 Seasons (plus Change of Seasons), the 5 Mountains of Daoism, the 5 Emperors related to the Planets and their relationships with the 5 Elements from ancient Greece (Earth, Fire, Water, Air and Aether) are listed here. These in turn are related to the 5 suits of the Tarot cards (Coins, Swords, Batons, Cups, and the Major Arcana suit) and they are also listed here in J. Coins, Batons and Cups (or Diamonds, Clubs and Hearts) then correspond with the 3 Mahjong suits (Coins 筒(饼), Strings of Coins 索(條) and *Wan* or 10,000 coins 万(萬)) and they too are listed in J. [Volume 4]
K The 7 Days of the Week (English Days, Roman Days, Japanese and Chinese Days) and the 7 Chinese Luminaries (Sun, Moon, Mars, Mercury, Jupiter, Venus, Saturn (including the legendary Fú Xī 伏羲 as a sun god and Nǚ Wā 女娲 as a moon goddess). [Volume 2 and Volume 4]
L The 1st, 2nd, 3rd and 4th weeks of a month relating to the 28 Lunar Mansions, the 48 Flower Cards and the 5 Directional animals. [Volume 4]
M The 28 Chinese Lunar Mansions (or Asterisms) and their animals [Volume 4]

N   The 24 Chinese Solar Terms [Volume 4]
O   The 10 Heavenly Stems from China [Volume 4]
P   The 12 Earthly Branches and the 12 Zodiacal Animals from China [Volume 4]
Q   The 60 Year Cycle from the Chinese calendar with events in China from 1864/1865 to 1923/1924 [Volume 4]
R   Infinity, the Supreme Ultimate, the Two Forms (Yin and Yang) and the 4 Phenomena [Volume 4]
S   The 8 Trigrams or Bā Guà of Earlier Heaven [Volume 4]
T   The 8 Trigrams or Bā Guà of Later Heaven [Volume 4]
U   The 8 Martial Arts Trigrams or Bā Guà [Volume 4]
V   The 64 Hexagrams from the Yì Jīng 易经. [Volume 4]
W   The 9 Stars of the Northern Dipper [Volume 4]
X   The 6 Stars of the Southern Dipper [Volume 4]
Y   The Nine Sons of the Dragon plus 3 extras. [Volume 4]
Z   The 48 Japanese Flower Cards (Hanafuda) divided into 12 months and 48 weeks corresponding to the 24 Chinese Solar Terms and the 12 signs of the zodiac. [Volume 4]

Jupiter is the 6th planet (or Dwarf Planet) in E and its moon Amalthea is its 3rd moon. The drawing of Amalthea therefore has the number 6/3 in E. Similarly the drawing of Pandora, the 6th moon of the 7th planet (or Dwarf Planet) Saturn, has the number 7/6 in E.

Asteroid 216 Kleopatra is the 67th asteroid in this selection in F and its moon Cleoselene is the closest to the parent body and therefore is numbered 1. The drawing for Cleoselene thus has the number 67/1 and is in F. Asteroid 87 Sylvia is the 97th asteroid in this selection in F and its moon Romulus is the 2nd out from the parent body and is therefore numbered 2. The drawing for Romulus thus has the number 97/2 and is placed in F.

The Tarot suit of Cups (corresponding with the Greek element Water and the Chinese element Water) is number 5 in J. The Mahjong suit of *Wan* (10,000 coins) which also equates with the Chinese Water element is numbered 5 in J. Thus the joint drawing for both the 6 of *Wan* and the 6 of Cups is placed in J and has the Water element 5 printed on it with the corresponding playing card image of a heart appearing at the top left of the card. This enables it to be easily recognized.

The Major Arcana Tarot suit which largely reflects religious and spiritual ideas is equated with the Greek element Aether and with the Chinese element Metal, both of which are in J and the joint card has the Metal element 4 printed on it. The drawing for Major Arcana number 9 (The Hermit) is placed in J (with the Metal element 4 printed on it) and numbered 9 at the top left of the card next to the image of a tower which I have adopted as the symbol for Major Arcana.

Symbols representing each Series of Drawings 符号

The 5 Chinese Elements 五行

The 5 Ancient Greek Elements 古希腊五行

1 木
Wood

The Chinese Element Wood = Greek Air

Air 空气 **1** 木
Wood

The Greek Element Air = Chinese Wood

2 火
Fire

The Chinese Element Fire = Greek Fire

3 土
Earth

The Chinese Element Earth = Greek Earth

4 金
Metal

The Chinese Element Metal = Greek Aether (Spirit)

5 水
Water

The Chinese Element Water = Greek Water

Astronomical Pencil Drawings - John Oxenham Goodman

Chinese 4 Seasons & Change of Seasons 四季和四季变化

Chinese 5 Emperors represent 5 Planets 中国古代五帝五星关系

5 Mountains of Chinese Daoism 五岳真形图

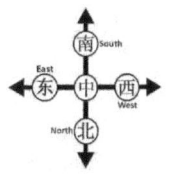

Animals of Chinese 4 Directions and Centre 五方动物

The 7 Days of the Week (English, Roman, Japanese and Chinese) 一周七天(英国,古罗马, 日本和中国)

The 7 Chinese Luminaries (Sun, Moon, and 5 Planets 中国七曜（日，月，五星

The 28 Chinese Lunar Mansions 二十八宿

The 24 Chinese Solar Terms 二十四节气

The 48 Japanese Flower Cards 四十八花札

The 12 Signs of the Zodiac 西方十二星座

The 10 Chinese Heavenly Stems 十天干

The 12 Chinese Earthly Branches and Yearly Animals 十二地支和十二生肖

1st Year 第一年
60th Year 第六十年

The Chinese 60 Year Cycle 六十甲子

Infinity, The Supreme Ultimate, The 2 Forms & The 4 Phenomena 无极，太极，两仪和四象

The 8 Trigrams (Bā Guà) of Earlier Heaven 先天八卦

The 8 Trigrams (Bā Guà) of Later Heaven 后天八卦

The 8 Martial Arts Trigrams (Bā Guà) 尹氏八卦掌

The 64 Hexagrams of the Yì Jīng 易经的六十四卦

The 9 Stars of the Northern Dipper 北斗九星(九皇大帝)

The 6 Stars of the Southern Dipper 南斗六星

The 9 Sons of the Dragon plus 3 = 12 龙生九子 + 3 = 12

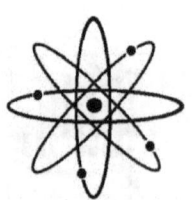

The 112 Modern Chemical Elements 一百一十二化学元素

The 88 Modern Constellations 八十八星座

The 10 Shang Dynasty Suns 商朝十太阳

Sun, Planets & Moons of the Solar System 古希腊太阳神和太阳系的行星和月球

The Planets and Moons of the Solar System 太阳系的行星和月球

A Selection of 120 Asteroids, their Moons and 4 Comets 小行星，月球和彗星

Chinese Chess (14 Different chess pieces) 中国象棋 (14 个不同的棋子)

Spades = Tarot Swords, Chinese Element Wood (Tens) 木=空气，塔罗牌的剑，黑桃

Clubs = Tarot Sticks (Bamboo), Chinese Element Fire (100s) 火，麻将的条，塔罗牌的棍，草花

Diamonds = Tarot Coins, Chinese Element Earth (Units) 土，麻将的筒，塔罗牌的硬币，方片

Guard Towers on Great Wall = Tarot Major Arcana, Chinese Element Metal (1,000s) 金 = 苍天，瞭望塔 = 塔罗牌的大秘仪

Hearts =Tarot Cups, Chinese Element Water (10,000s) 水，塔罗牌的杯，麻将的万，红桃

Spades = Tarot Swords, Chinese Element Wood (Tens) 木=空气，塔罗牌的剑，黑桃

Clubs = Tarot Sticks (Bamboo), Chinese Element Fire (100s) 火，麻将的条，塔罗牌的棍，草花

Diamonds = Tarot Coins, Chinese Element Earth (Units) 土，麻将的筒，塔罗牌的硬币，方片

Guard Towers on Great Wall = Tarot Major Arcana, Chinese Element Metal (1,000s) 金 = 苍天，瞭望塔 = 塔罗牌的大秘仪

Hearts = Tarot Cups, Chinese Element Water (10,000s) 水，塔罗牌的杯，麻将的万，红桃

The 5 Chinese Elements (Wood, Fire, Earth, Metal, Water) as Suits in Composite Tarot/Mahjong Cards and their Relationship to the 5 Ancient Greek Elements (Air, Fire, Earth, Aether/Spirit, Water), the 5 Tarot Suits (Swords, Batons, Coins, Major Arcana, Cups) and the 3 Mahjong Suits (Sticks 条, Circles 筒, Wan 万)

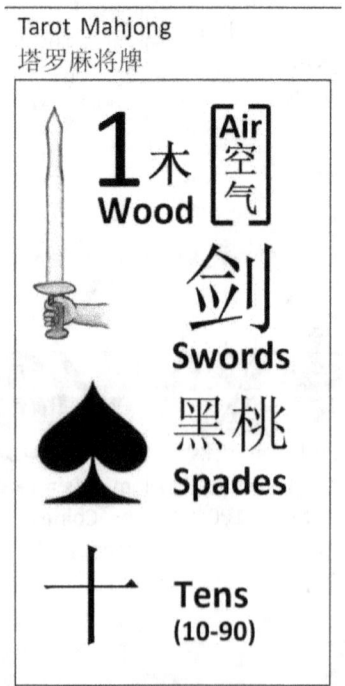

Tarot Mahjong
塔罗麻将牌

3 土
Earth

硬币 Coins

筒 Circles

◇ 方片 Diamonds

一位数 Units (1-9)

Tarot Mahjong
塔罗麻将牌

4 金
Metal
(Aether/Spirit)

长城岗搂
Great Wall Guard Tower

大秘仪
Major Arcana

精神世界   苍天
Spirits of the Aether
Beyond the Material World

千 Thousands
(1,000-9,000)

Tarot Mahjong
塔罗麻将牌

5 水
Water

杯 Cups

 红桃 Hearts

 Wàn
Ten Thousands
(10,000-90,000)

The Five Elements 五行

In the Generating arrangement each Chinese element creates or generates the element which follows it.

1. 木 Mù Wood feeds Fire.
2. 火 Huǒ Fire creates Earth (in the form of ash).
3. 土 Tǔ Earth bears Metal (which is mined from the earth).
4. 金 Jīn Metal collects water (which can fall as rain into metal vessels).
5. 水 Shuǐ Water nourishes Wood (which comes from trees)

The 5 Elements 五行 Wǔ Xíng are related to the Four Seasons plus Change of Seasons as well as to the Four Directions plus Centre.

3 The 5 Chinese Elements
中国五行

土 Earth 3土 Earth

Chemical Elements
27, 28, 52

中
Centre

Planets & Moons
3

Tarot
21

季节变化 Change of Seasons

4 The 5 Chinese Elements
中国五行

金 Metal 4金 Metal

西
W

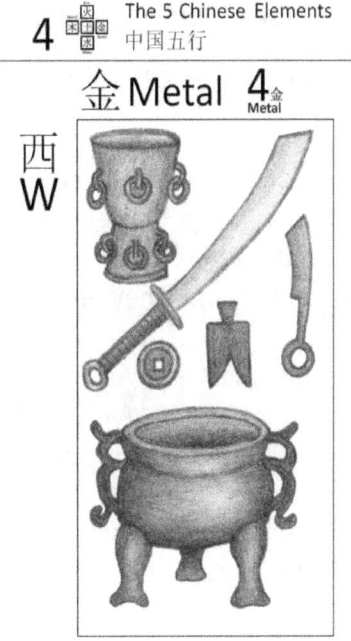

秋 Autumn

5 The 5 Chinese Elements
中国五行

水 Water 5水 Water

北
N

Tarot
12

冬 Winter

The Four Greek Elements and Aether (or Spirit) 古希腊五行

The four ancient Greek elements, devised by the Greek philosopher Empedocles in the 5th century BC, were the basic building blocks of matter and a fifth element Aether was the domain of the spirits in the non-material world. They relate to the Five Chinese Elements as follows:

Chinese Element   **Ancient Greek Element**
1. 木 Mù Wood   Air 空气 Kōng Qì
2. 火 Huǒ Fire   Fire 火 Huǒ
3. 土 Tǔ Earth   Earth 土 Tǔ
4. 金 Jīn Metal   Aether (Spirit) 苍天 Cāng Tiān (精神世界)
5. 水 Shuǐ Water   Water 水 Shuǐ

Sylph of the Air 空气精灵

Salamander of the Fire 火蜥蜴

3 The 5 Ancient Greek Elements 古希腊五行

Chemical Elements 27, 28, 52

Earth 土 ▽ 3 土 Earth

中 Centre

Planets & Moons 3

Tarot 闆 21

Gnome of the Earth 地精

4 The 5 Ancient Greek Elements 古希腊五行

Aether 苍天 (Spirit) 4 金 Metal

西 W

Beyond the Material World

Spirits of the Aether
精神世界

5 The 5 Ancient Greek Elements 古希腊五行

Water 水 ▽ 5 水 Water

北 N

Tarot 闆 12

Nymph of the Water
水中仙女

The 4 Seasons and the Change of Seasons
四季和季节变化

The Four Seasons 四季 (Sì Jì) and the Change of Seasons 季节变化 (Jì Jié Biàn Huà) are related to the five Chinese elements and the five directions.

Season
1. 春(Chūn) Spring
2. 夏(Xià) Summer
3. 季节变化 Change of Seasons
4. 秋(Qiū) Autumn
5. 冬(Dōng) Winter

Element
木 Mù Wood
火 Huǒ Fire
土 Tǔ Earth
金 Jīn Metal
水 Shuǐ Water

Direction
东(Dōng) East
南(Nán) South
中(Zhōng) Centre
西(Xī) West
北(Běi) North

1 The 4 Seasons and Change of Seasons 四季变化
春 Spring 1 木 Wood
东 E
Tarot ♠16,17

2 The 4 Seasons and Change of Seasons 四季变化
夏 Summer 2 火 Fire
南 S
Tarot ♣16,17

Běidàihé Beach, Qínhuáng-dǎo City, Héběi 秦皇岛市北戴河

3 — The 4 Seasons and Change of Seasons 四季变化

季节变化
Change of Seasons **3** 土 Earth

中 Centre

4 — The 4 Seasons and Change of Seasons 四季变化

秋 Autumn **4** 金 Metal

西 W

Tarot
♞ 24,25

5 — The 4 Seasons and Change of Seasons 四季变化

冬 Winter **5** 水 Water

北 N

Planets & Moons
7/29

Tarot
♡ 16,17

The 5 Emperors related to the 5 Planets
中国古代五帝五星关系

The Five Emperors are traditionally thought to have reigned from 2952 BC to 2437 BC. Many of their deeds lack historical verification and are regarded as mythical. They are related to the 5 elements and 5 directions and their relationship to the 5 planets is as follows:

The 5 Emperors
1 Tài Hào 太昊 (Fú Xī 伏羲)
2 Yán Dì 炎帝 (Shén Nóng 神农)
3 Huáng Dì 黄帝 (Yellow Emperor)
4 Shào Hào 少昊

5 Zhuān Xū 颛顼

The 5 Planets
木星 Mù Xīng Wood Star (Jupiter)
火星 Huǒ Xīng Fire Star (Mars)
土星 Tǔ Xīng Earth Star (Saturn)
金星 Jīn Xīng Gold (Metal) Star (Venus)
水星 Shuǐ Xīng Water Star (Mercury)

Emperor Tài Hào, Earthly Representative of Planet Suì Xīng 岁星 or Mù Xīng 木星 (Jupiter) with his wife Nǚ Wā 女娲. Tài Hào is also known as Cāng Dì 苍帝 and Fú Xī 伏羲.

Emperor Yán Dì Earthly Representative of Planet Yíng Huò 荧惑 or Huǒ Xīng 火星 (Mars). Yán Dì is also known as Shén Nóng 神农 (the Divine Farmer) and Zhù Róng 祝融.

3 ● The 5 Emperors related to the 5 Planets 中国古代五帝 五星关系

Huáng Dì 黄帝 **3** 土 Earth

中 Centre

Planets & Moons **7**

黄帝 The Yellow Emperor

Emperor Huáng Dì, Earthly Representative of Planet Zhèn Xīng 镇星, Tián Xīng 填星 or Tǔ Xīng 土星 (Saturn).

4 ● The 5 Emperors related to the 5 Planets 中国古代五帝 五星关系

Shào Hào 少昊 **4** 金 Metal

西 W

Planets & Moons **2**

白帝 The White Emperor

朱宣 Zhū Xuān
长庚 Cháng Gēng
启明 Qī Ming

Emperor Shào Hào, Earthly Representative of Planet Tài Bái Jīn Xīng 太白金星 (Venus). Shào Hào is also known as Zhū Xuān 朱宣, Qī Míng 启明 and Cháng Gēng 长庚.

5 ● The 5 Emperors related to the 5 Planets 中国古代五帝 五星关系

Zhuān Xū 颛顼 **5** 水 Water

北 N

Planets & Moons **1**

黑帝 The Black Emperor

Emperor Zhuān Xū (Xuán Dì 玄帝), Earthly Representative of Planet Chén Xīng 辰星 or Shuǐ Xīng 水星 (Mercury).

The 5 Mountains of Daoism and their Diagrams 五岳真形图

Some say that the ancient diagrams were an attempt to depict the shapes of the mountains while others speculate that they were plans of the routes taken by climbers. The diagrams or sketches were confirmed more than 2000 years ago during the Qin Dynasty or not later than the Han Dynasty. Mountains were worshipped during that period and the number five corresponds to the five elements, five directions and five colours. Tái Shān was climbed by 72 emperors.

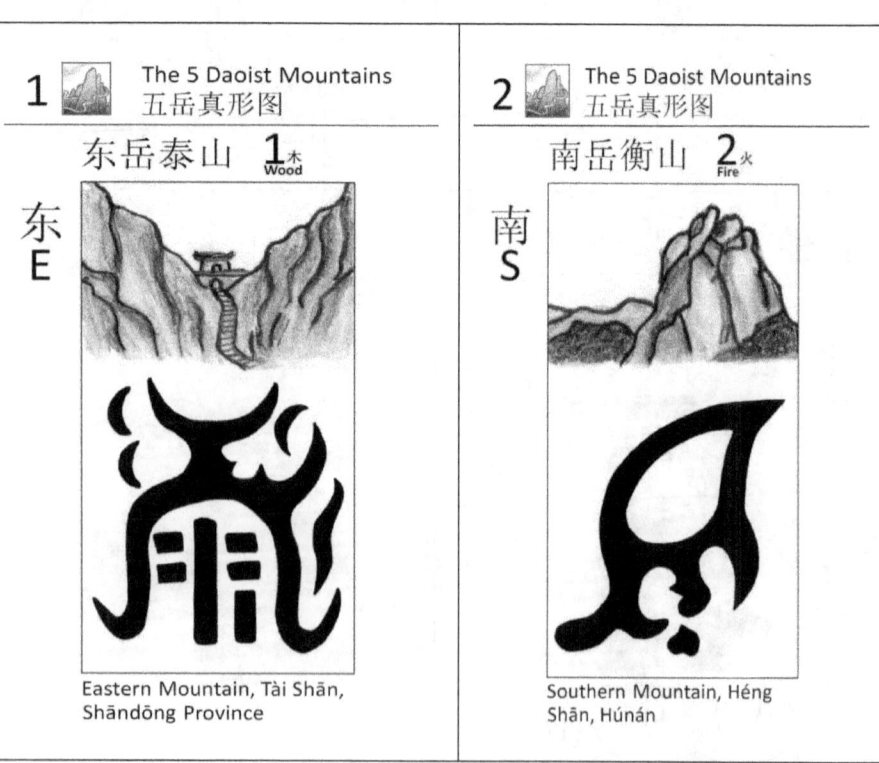

3	The 5 Daoist Mountains 五岳真形图

中岳嵩山 **3** 土 Earth

中 Centre

Central Mountain, Sōng Shān, Hénán Province

4	The 5 Daoist Mountains 五岳真形图

西岳华山 **4** 金 Metal

西 W

Western Mountain, Huá Shān, Shaǎn Xī Province

5	The 5 Daoist Mountains 五岳真形图

北岳恒山 **5** 水 Water

北 N

Northern Mountain, Héng Shān, Shānxī Province

The Animals of the 5 Directions 五方动物

The Animals of the 5 Directions 五方动物 Wǔ Fāng Dòng Wù are related to the 5 elements and 5 directions. Animal 3 is the Yellow Dragon of the Centre which represents the Earth and the Emperor. Animals 1, 2, 4 and 5 are in the sky and each of them encompasses 7 of the 28 Chinese Lunar Mansions and one of the 4 weeks of a month.

Directional Animal	Element	Lunar Mansions	Weeks
1 Azure Dragon of the East 东方青龙	Wood	1-7	Week 1
2 Red Bird of the South 南方朱雀	Fire	22-28	Week 4
3 Yellow Dragon of the Centre 中央黄龙	Earth		
4 White Tiger of the West 西方白虎	Metal	15-21	Week 3
5 Black Warrior of the North 北方玄武	Water	8-14	Week 2

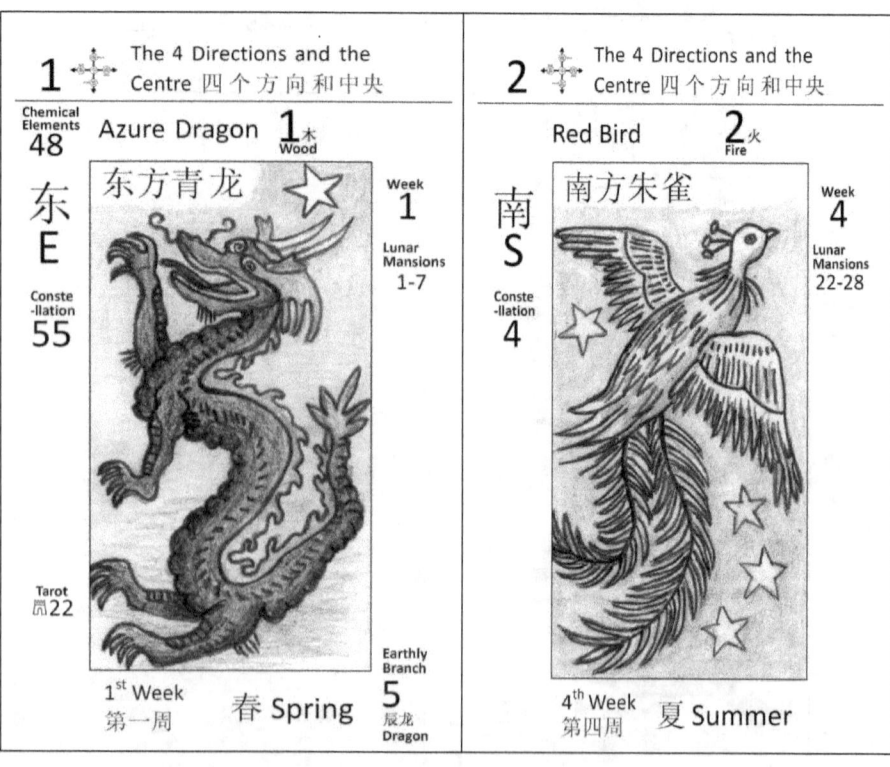

Card 3: Yellow Dragon

3 ✥ The 4 Directions and the Centre 四个方向和中央

Chemical Elements **48**

Yellow Dragon **3** 土 Earth

中央黄龙

中 Centre

Constellation **55**

Lunar Mansion **1, 2**

Tarot 🀄 **22**

季节变化 Change of Seasons

Earthly Branch **5** 辰龙 Dragon

Card 4: White Tiger

4 ✥ The 4 Directions and the Centre 四个方向和中央

White Tiger **4** 金 Metal

西方白虎

西 W

Week **3**

Lunar Mansions **6, 15-21**

3rd Week 第三周

秋 Autumn

Earthly Branch **3** 寅虎 Tiger

Card 5: Black Warrior

5 ✥ The 4 Directions and the Centre 四个方向和中央

Black Warrior **5** 水 Water

北方玄武

北 N

Constellation **8**

Week **2**

Lunar Mansions **8-14, 27**

Tarot ♡ **24**

龟和蛇 Tortoise and Snake

2nd Week 第二周

冬 Winter

Earthly Branch **6** 巳蛇 Snake

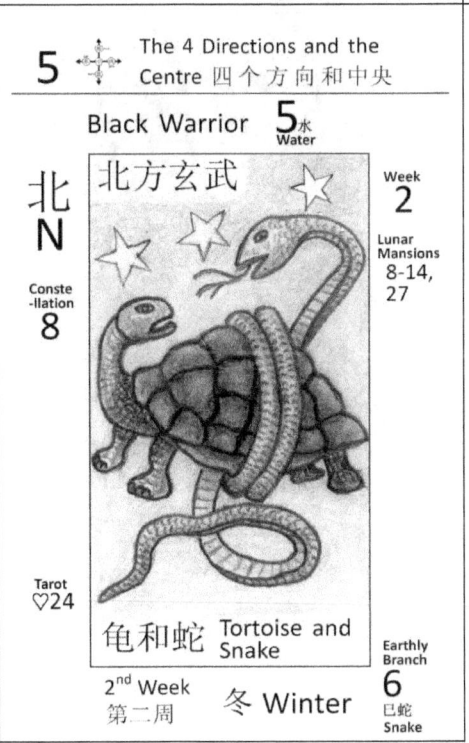

The Old English Days of the Week 古英文的星期

The old English or Anglo Saxon days of the week are named after the Teutonic, Germanic or Norse gods and goddesses except for Saturday which is named after Saturn the Roman God of Agriculture. Some of these deities share common attributes with the Roman gods. For example both Tyr and Mars are gods of war; Thor and Jupiter both use thunder or lightning as a weapon; Frigg and Venus are both goddesses of love. However, the Roman sun god Sol is masculine while his English counterpart is feminine. Likewise the moon was a woman in ancient Rome but a man in old England.

The Roman Days of the Week 古罗马的星期

The Roman days of the week are named after the Roman gods and goddesses some of which share common attributes with their English counterparts. For example both Mars and Tyr are gods of war while both Jupiter and Thor employ lightning as a weapon. Venus and Frigg are both goddesses of love and Saturn is the same in both systems. However, the sun was a god in ancient Rome but a goddess in old England while the moon goddess in Rome became a god in old England. The names of the days of the week in Italy, France, Spain and Portugal are derived from the old Roman names.

1 The Roman Days of the Week 古罗马的星期 Chemical Elements **34** **Dies Lunae** (Monday) 星期一 **5** 水 Water Day **1** Mon Planets & Moons **3/1**  Tarot 2, 18 Day of Luna, Roman Moon Goddess 古罗马月亮女神	**2** The Roman Days of the Week 古罗马的星期 **Dies Martis** (Tuesday) 星期二 **2** 火 Fire Day **2** Tue Planets & Moons **4** 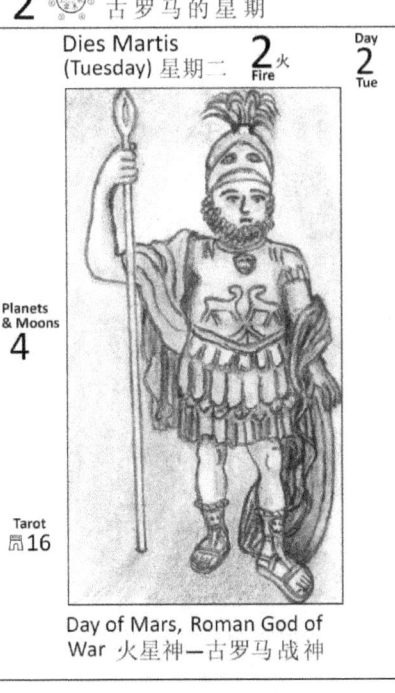 Tarot 16 Day of Mars, Roman God of War 火星神—古罗马战神
3 The Roman Days of the Week 古罗马的星期 Chemical Elements **80** **Dies Mercurii** (Wednesday) 星期三 **5** 水 Water Day **3** Wed Planets & Moons **1** 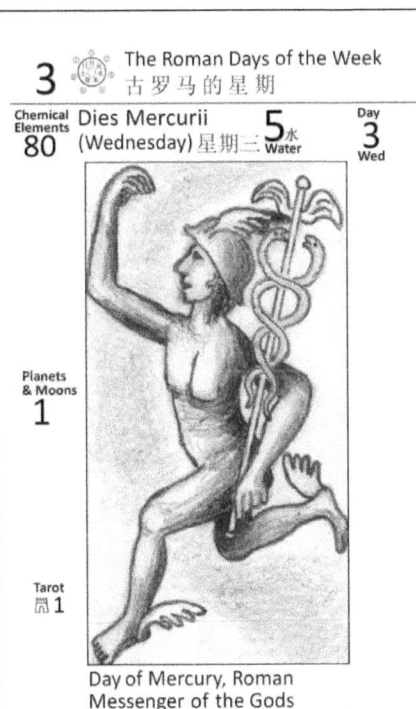 Tarot 1 Day of Mercury, Roman Messenger of the Gods 水星神墨丘利—众神的使者	**4** The Roman Days of the Week 古罗马的星期 **Dies Jovis** (Thursday) 星期四 **1** 木 Wood Day **4** Thu Planets & Moons **6** 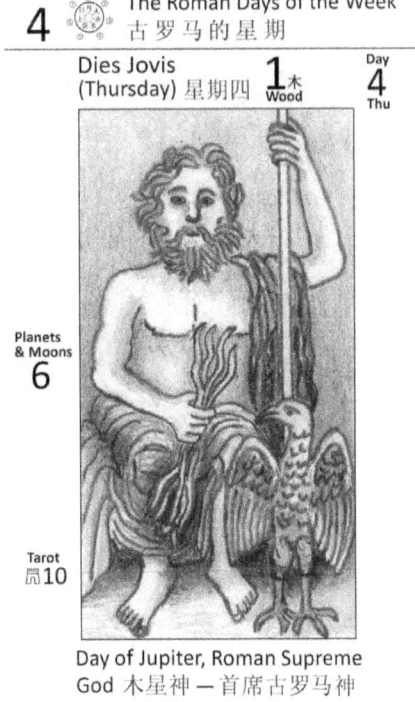 Tarot 10 Day of Jupiter, Roman Supreme God 木星神—首席古罗马神

5 — The Roman Days of the Week 古罗马的星期

Chemical Elements 15
Dies Veneris (Friday) 星期五
4 金 Metal
Day 5 Fri

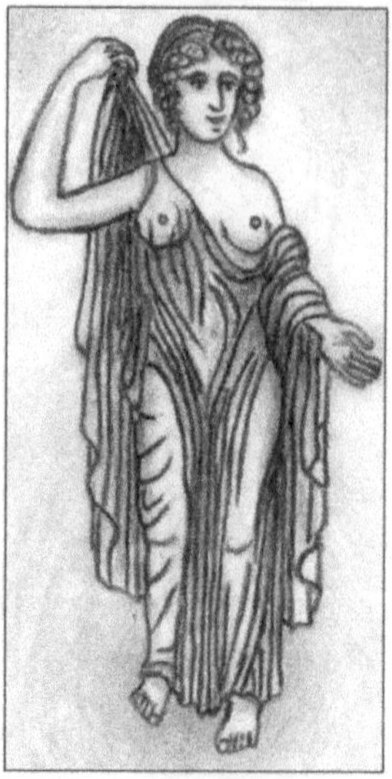

Planets & Moons 2

Tarot 3

Day of Venus, Roman Goddess of Love 金星神维纳斯—爱女神

6 — The Roman Days of the Week 古罗马的星期

Chemical Elements 52
Dies Saturni (Saturday) 星期六
3 土 Earth
Day 6 Sat

Planets & Moons 7

Tarot 21

Day of Saturn, Roman God of Agriculture 土星神—罗马农业神

7 — The Roman Days of the Week 古罗马的星期

Chemical Elements 2, 42, 60, 89, 91
Dies Solis (Sunday) 星期日
2 火 Fire
Day 7 Sun

Shang Suns 3

Sun, Planets & Moons 0

Tarot 19

Day of Sol, Roman Sun God
古罗马太阳神

The Japanese and Chinese Days of the Week 日本中国的星期

The modern Japanese days of the week, like the Roman days of the week, are named after the sun, moon and five planets and are in the same order. The 5 planets are themselves named after the 5 Chinese elements. The modern Chinese days of the week are numbered 1 to 6 plus Sunday.

The 5 Elements	The 5 Planets	Days
2. 火 Huǒ Fire	火星 Huǒ Xīng Fire Star (Mars)	Tuesday
5. 水 Shuǐ Water	水星 Shuǐ Xīng Water Star (Mercury)	Wednesday
1. 木 Mù Wood	木星 Mù Xīng Wood Star (Jupiter)	Thursday
4. 金 Jīn Metal	金星 Jīn Xīng Metal Star (Venus)	Friday
3. 土 Tǔ Earth	土星 Tǔ Xīng Earth Star (Saturn)	Saturday

The 7 Chinese Luminaries 中国七曜

These are Chinese representations in art of the gods of the Sun, Moon and five planets Mercury, Venus, Mars, Jupiter and Saturn. They are very different from their Roman equivalents in both name and appearance. Several different images of sun gods and moon goddesses appear here, including a 7000 year old stone engraving from Zǐ Guī 姊归 in Hunan Province, an ancient rock engraving from Hè Lán Shān 贺兰山 in Ningxia as well as Han Dynasty depictions of Fú Xī 伏羲 as a sun god and his wife Nǔ Wā 女娲 as a moon goddess. An image of moon goddess Cháng É 嫦娥 has also been added and the old white haired Tài Bái Jīn Xīng 太白金星 now represents Venus. In the earliest times planet Venus was a goddess who played the *pípa* 琵琶 and wore a long yellow skirt, and she is shown here too.

(1) 1 — The 7 Chinese Luminaries 中国七曜（日，月，五星） Earth's Moon 地球之月球

Chemical Elements: 34, 59
The Moon 月亮
5 Water 水
Day 1 Mon
Planets & Moons: 3/1
Tarot: 2, 18

Cháng É, Goddess of the Moon 嫦娥—月亮女神

(2) 1 — The 7 Chinese Luminaries 中国七曜（日，月，五星） Earth's Moon 地球之月球

Chemical Elements: 34, 59
Moon Goddess Nǚ Wā 月神女娲
5 Water 水
Day 1 Mon
Planets & Moons: 3/1
Tarot: 2, 18

Nǚ Wā, with bird's body and wings, carries the moon and its Jade Rabbit, Toad and Cassia Tree 人首鸟身，背负着月轮，月轮里有玉兔，蟾蜍和桂树

(3) 1 — The 7 Chinese Luminaries 中国七曜（日，月，五星） Earth's Moon 地球之月球

Chemical Elements: 34, 59
Moon Goddess Nǚ Wā 月神女娲
5 Water 水
Day 1 Mon
Planets & Moons: 3/1
Tarot: 2, 18

Nǚ Wā, with dragon legs and tail, supports the Moon and its Toad and Cassia Tree 人身龙尾，龙尾上有两只爪，托举着月亮，月亮里有蟾蜍和桂树

2 — The 7 Chinese Luminaries 中国七曜（日，月，五星）

荧惑 **Yíng Huò Mars**
2 Fire 火
Day 2 Tue
Planets & Moons: 4
Tarot: 16, 20

火星 Huǒ Xīng Fire Planet

3 ☀☽ The 7 Chinese Luminaries 中国七曜（日，月，五星） 辰星 Chén Xīng **5** 水 Day **3** Wed Mercury Water Planets & Moons **1** Tarot 🁢 1, 12 水星 Shuǐ Xīng Water Planet 	**4** ☀☽ The 7 Chinese Luminaries 中国七曜（日，月，五星） 岁星 Suì Xīng **1** 木 Day **4** Thu Jupiter Wood Planets & Moons **6** Tarot 🁢 0, 10 木星 Mù Xīng Wood Planet 
(1) **5** ☀☽ The 7 Chinese Luminaries 中国七曜（日，月，五星） Tài Bái Jīn Xīng 太白金星(Venus) **4** 金 Day **5** Fri Metal 西 W Planets & Moons **2** Tarot 🁢 3 Tài Bái Jīn Xīng 太白金星, The Great White Golden Planet (Venus) 	(2) **5** ☀☽ The 7 Chinese Luminaries 中国七曜（日，月，五星） Tài Bái Jīn Xīng 太白金星(Venus) **4** 金 Day **5** Fri Metal 西 W Planets & Moons **2** Tarot 🁢 3 Goddess of Planet Venus before Ming Dynasty with *pípa*, cockscomb and long yellow skirt 明朝前太白金星是女神。 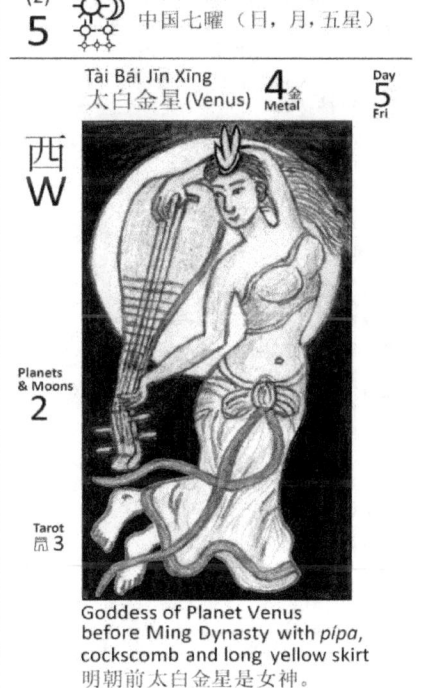

6 ☉☽ ✦✦✦ The 7 Chinese Luminaries 中国七曜（日，月，五星）	**(1) 7** ☉☽ ✦✦✦ The 7 Chinese Luminaries 中国七曜（日，月，五星） The Eastern Han Dynasty Sun God 东汉日神
镇星 Zhèn Xīng Saturn **3** 土 Earth Day **6** Sat Planets & Moons **7** Tarot 🂡 21  土星 Tǔ Xīng Earth Planet (also known as Tián Xīng 填星)	Chemical Elements 2, 42, 60, 89, 91 Sun God Fú Xī 日神伏羲 **2** 火 Fire Day **7** Sun Shang Suns **3** Sun, Planets & Moons 0, 6/66 Asteroids & Moons, Comets 1, 10, 93 Tarot 🂡 19 Fú Xī, with bird's body and wings, carries the Sun and its Sun Bird 伏羲，人首鸟身，背负带有金乌的日轮
(2) 7 ☉☽ ✦✦✦ The 7 Chinese Luminaries 中国七曜（日，月，五星） The Eastern Han Dynasty Sun God 东汉日神	**(3) 7** ☉☽ ✦✦✦ The 7 Chinese Luminaries 中国七曜（日，月，五星）
Chemical Elements 2, 42, 60, 89, 91 Sun God Fú Xī 日神伏羲 **2** 火 Fire Day **7** Sun Shang Suns **3** Sun, Planets & Moons 0, 6/66 Asteroids & Moons, Comets 1, 10, 93 Tarot 🂡 19 Fú Xī, with dragon legs and tail, supports the Sun and its Sun Bird 伏羲，人身龙尾，龙尾上有两只爪，托举着太阳及太阳鸟	Chemical Elements 2, 42, 60, 89, 91 Sun Man, Húnán 湖南秭归太阳人 **2** 火 Fire Day **7** Sun Shang Suns **3** Sun, Planets & Moons 0, 6/66 Asteroids & Moons, Comets 1, 10, 93 Tarot 🂡 19 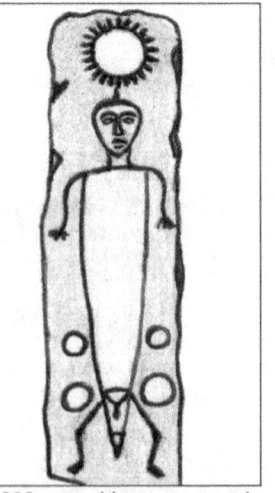 7,000 year old stone engraving of Sun Man at Zǐ Guī, Húnán 湖南姊归出土的 7000 年前的太阳人石刻。

Sun God rock engraving at Hèlán Shān, Níngxià, 3,000 to 10,000 years old
距今约 3000–10000 年前的宁夏贺兰山太阳神岩画

The 28 Chinese Lunar Mansions or Zodiacal Asterisms 二十八宿

The 28 Lunar Mansions (二十八宿 Èr Shí Bā Xiù) are the Chinese constellations of the zodiac. However, they are based on the moon, whereas the 12 Western zodiacal constellations are solar. These 28 lunar mansions or asterisms are divided into four groups: East, North, West and South according to their place in the sky. Each of the four groups contains seven asterisms and is distinguished by an element, a season, an animal, a color and a week of the month. Each lunar mansion is assigned a day of the week.

Astronomical Pencil Drawings - John Oxenham Goodman

13 ★ The 28 Lunar Mansions 二十八宿

猪 Pig　5 水 Water　Day 2 Tue　Week 2　Lunar Mansion 13

北 N　室火猪

Asteroids & Moons, Comets
13, 47

Earthly Branch
12 亥猪 Pig

2nd Week, Tuesday
第二周星期二

14 ★ The 28 Lunar Mansions 二十八宿

猰貐 Yà Yǔ Monster　5 水 Water　Day 3 Wed　Week 2　Lunar Mansion 14

北 N　壁水貐

2nd Week, Wednesday
第二周星期三

14/2 ★ The 28 Lunar Mansions 二十八宿

猰貐 Yà Yǔ Monster　5 水 Water　Day 3 Wed　Week 2　Lunar Mansions 14

北 N　壁水貐

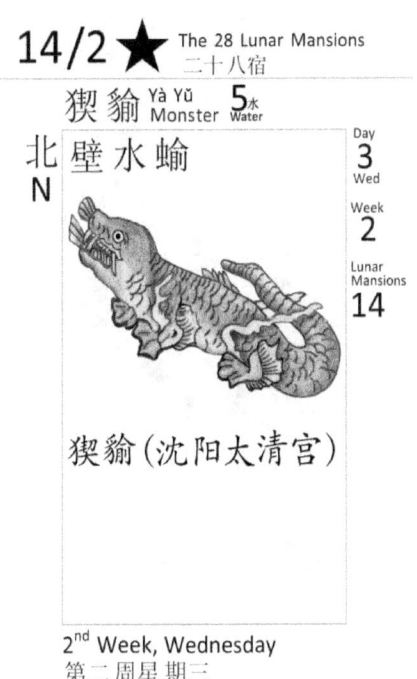

猰貐（沈阳太清宫）

2nd Week, Wednesday
第二周星期三

15 ★ The 28 Lunar Mansions 二十八宿

Chemical Elements 74　狼 Wolf　4 金 Metal　Day 4 Thu　Week 3　Lunar Mansion 15

西 W　奎木狼

Conste-llation 57

Planets & Moons
6/54, 7/29, 34, 39, 40, 46, 48, 57

Asteroids & Moons, Comets
97/1, 97/2

3rd Week, Thursday
第三周星期四

20 ★ The 28 Lunar Mansions 二十八宿

猴子 Monkey **4** 金 Metal
西 W
觜火猴

Day **2** Tue
Week **3**
Lunar Mansion **20**

3rd Week, Tuesday
第三周星期二

Earthly Branch **9** 申猴 Monkey

21 ★ The 28 Lunar Mansions 二十八宿

长臂猿 Gibbon **4** 金 Metal
西 W
参水猿

Day **3** Wed
Week **3**
Lunar Mansion **21**

3rd Week, Wednesday
第三周星期三

22 ★ The 28 Lunar Mansions 二十八宿

犴犴 Bì Àn (Dragon Dog) **2** 火 Fire
南 S
井木犴

Day **4** Thu
Week **4**
Lunar Mansion **22**

Tarot 罔 11

4th Week, Thursday
第四周星期四

23 ★ The 28 Lunar Mansions 二十八宿

羊 Goat **2** 火 Fire
南 S
鬼金羊

Day **5** Fri
Week **4**
Lunar Mansion **23**

Conste-llation 24,79

Planets & Moons 6/2, 6/3, 6/24, 6/25

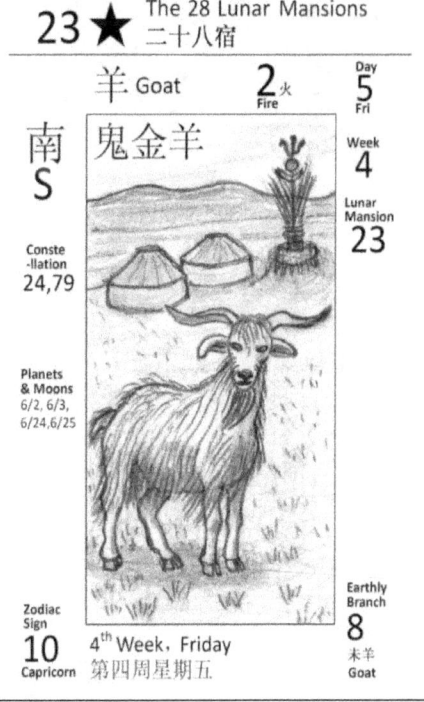

Zodiac Sign **10** Capricorn

4th Week, Friday
第四周星期五

Earthly Branch **8** 未羊 Goat

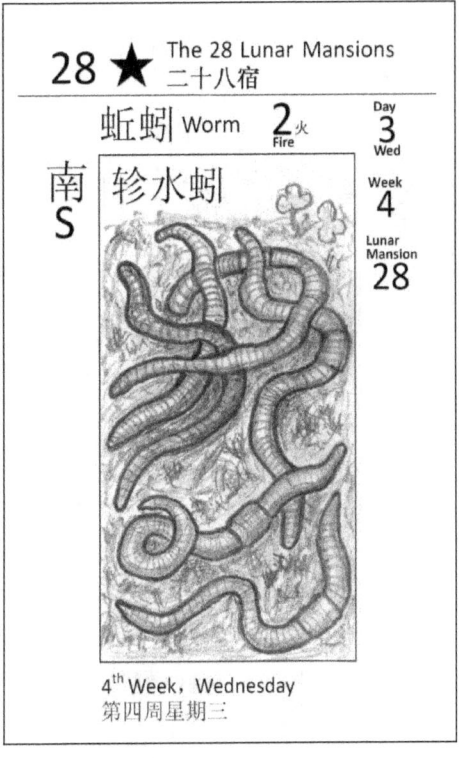

The 24 Solar Terms 二十四节气

The 24 Solar Terms (二十四节气 Èr Shí Sì Jié Qì) divide the year into 24 periods which are aligned with the seasons so that planting and harvesting of crops is commenced at the correct time. The 24 Solar Terms are arranged around the ecliptic 15 degrees apart and each solar term represents a period of about 15 days. Each solar term can be divided into 3 *Hou* 候 which are mostly 5 days in length and thus a year comprises 72 *Hou*. Four of the solar terms begin at the winter and summer solstices and the spring and autumn equinoxes and these were in use more than 3000 years ago during the Shang Dynasty. The current list of 24 appeared during the Western Han Dynasty (206 BC-23 AD). The solar terms correspond fairly closely with the 12 signs of the Western Zodiac. Each zodiac sign usually begins or ends within one day of the start of a major solar term.

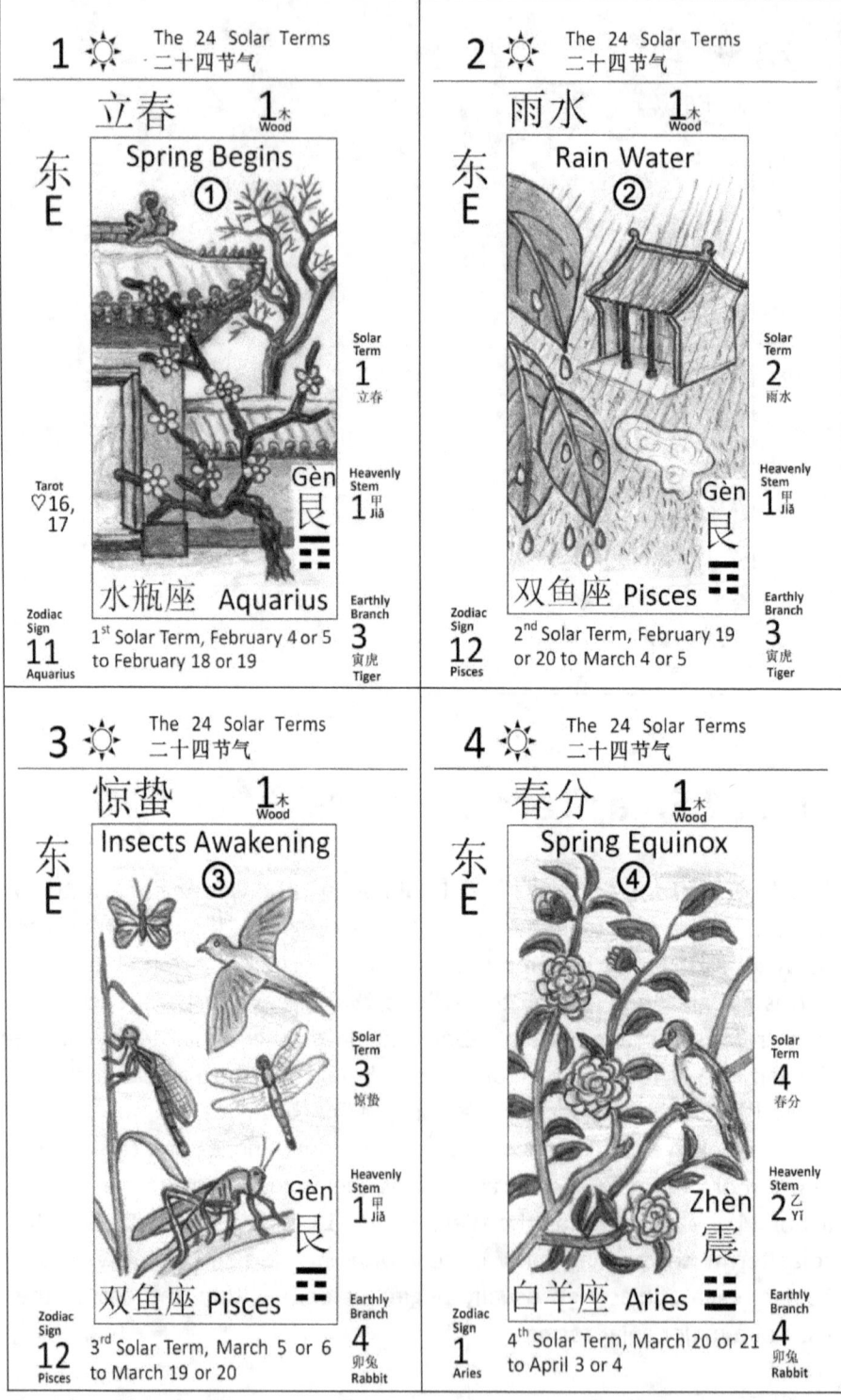

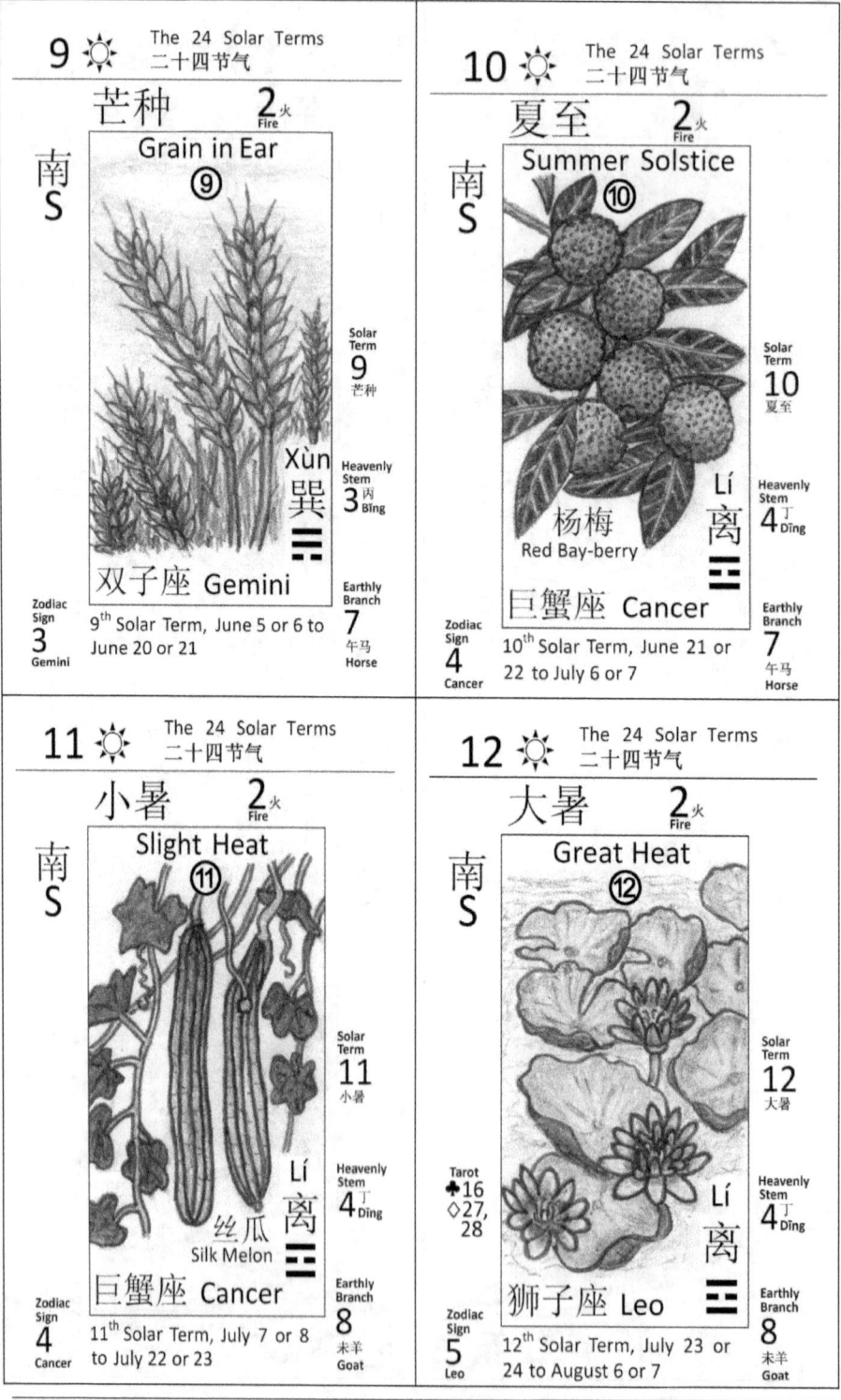

The 12 Months of the Japanese Flower Cards 花札 (Hanafuda)

Hanafuda (flower cards) have four cards for each month in a 12 month period and they represent the 4 weeks of each month. A different flower or tree is depicted each month and there are birds, animals, as well as the sun and moon which are relatable to items in other series. The cards can be related to the 24 solar terms, the 12 signs of the zodiac, the 8 *bā guà* or trigrams together with the 10 heavenly stems and the 12 earthly branches. Four of the five elements are relatable based on the 24 solar terms with the third element Earth which represents Change of Seasons being unassigned. Planets and Moons 7/8 Janus, 4 Mars, 2 Venus and Asteroids and Moons 48 Juno are relatable because they gave their names to 4 of the months. The lightning strike associated with Saturn's moon Farbauti 7/50 and Thor's lightning (Day 4 Wednesday, Chemical Elements 90), as well as the trigram Zhen 震 Thunder 雷, can be related to Hanafuda 11/4.

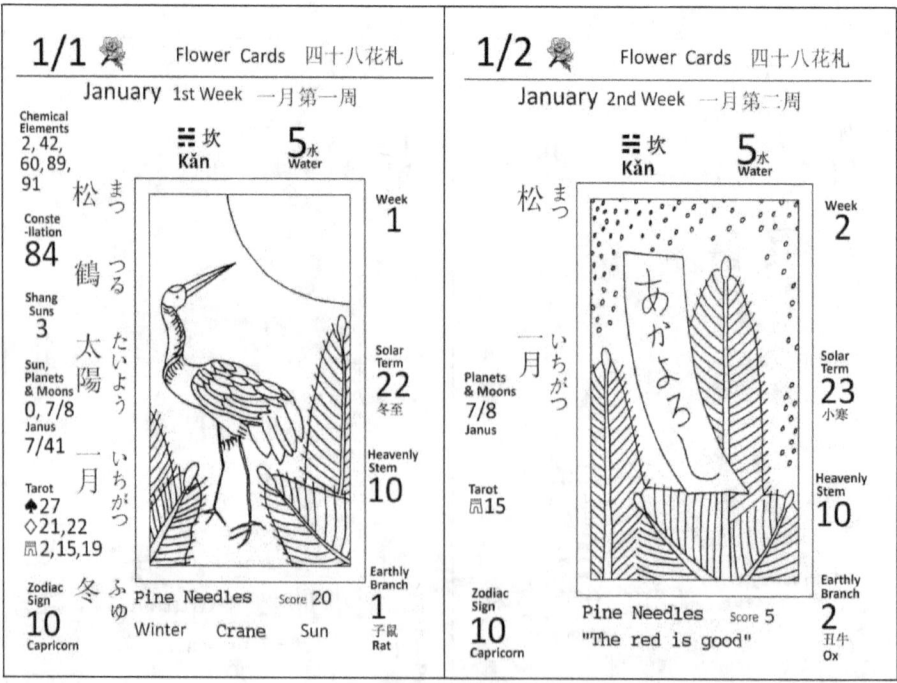

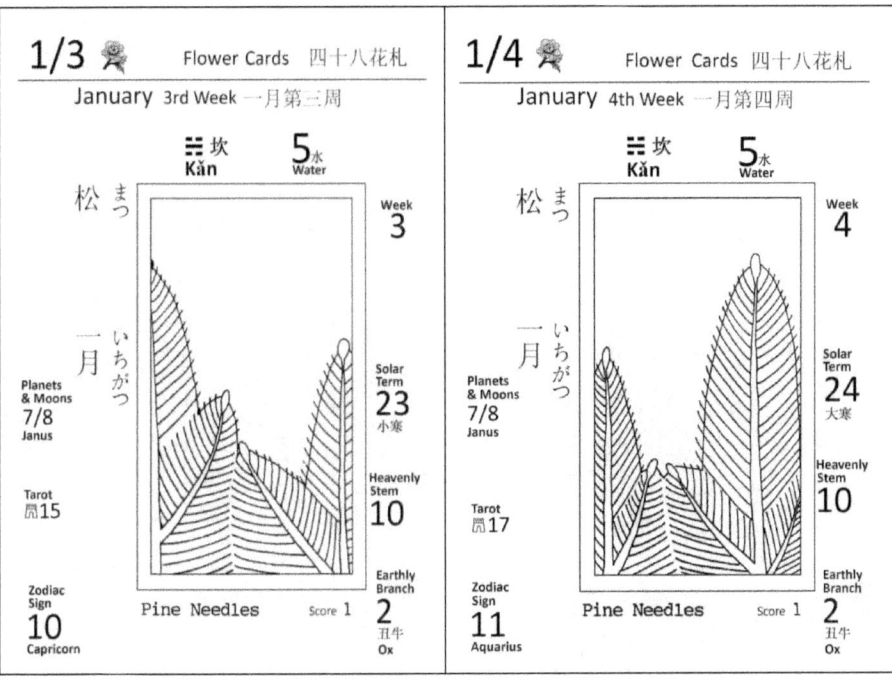

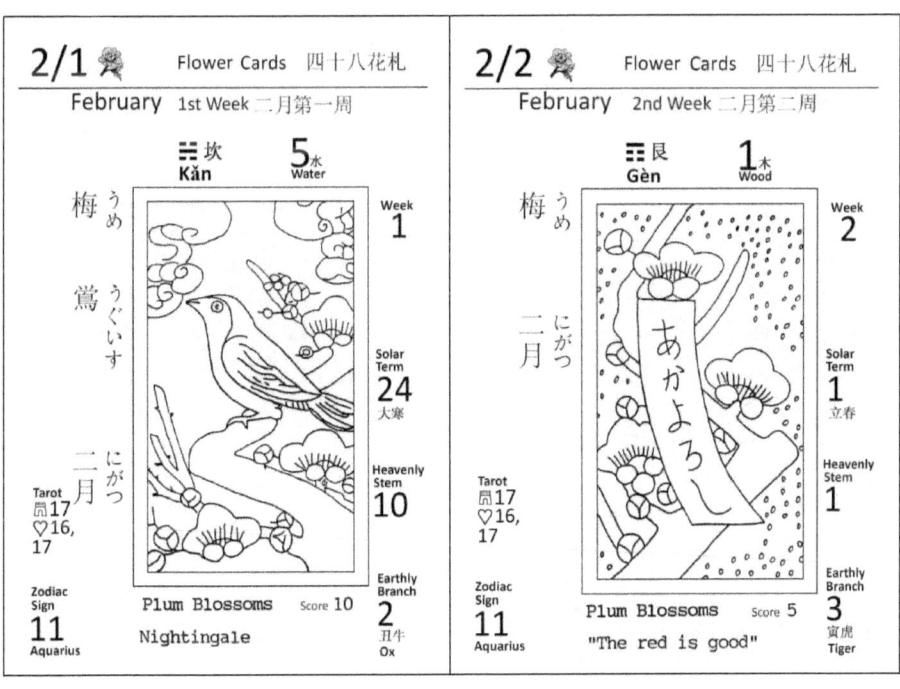

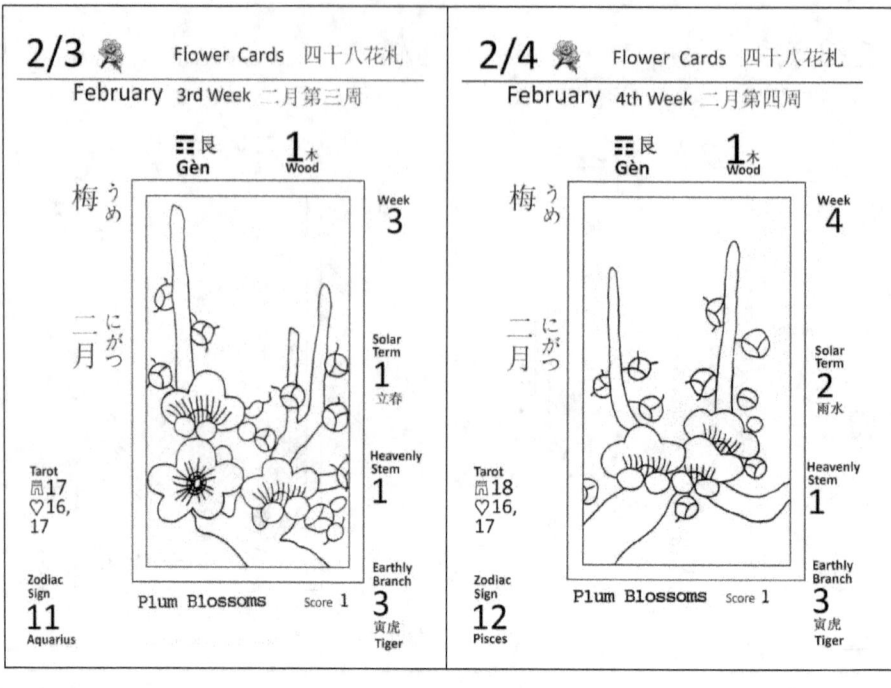

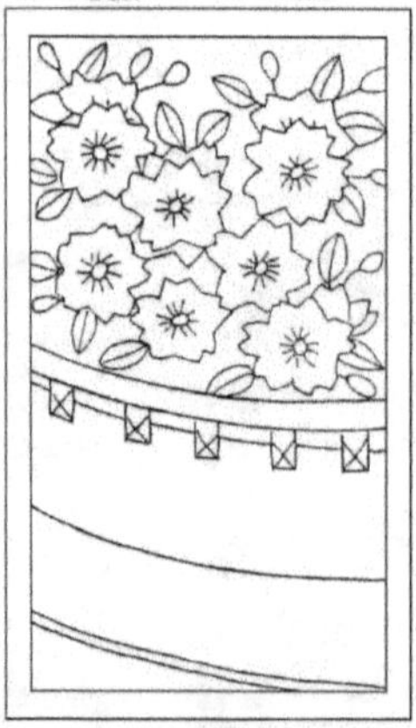

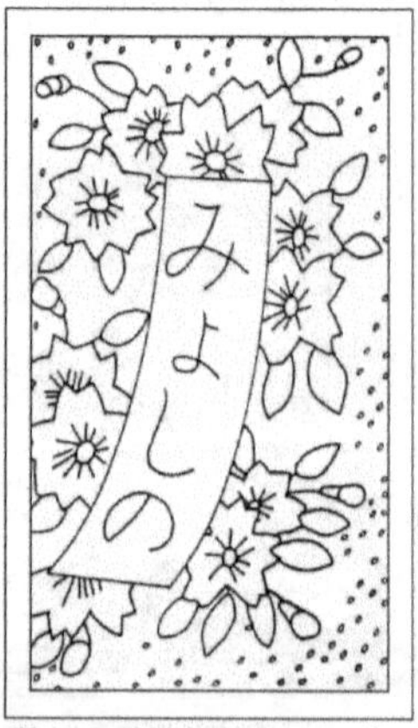

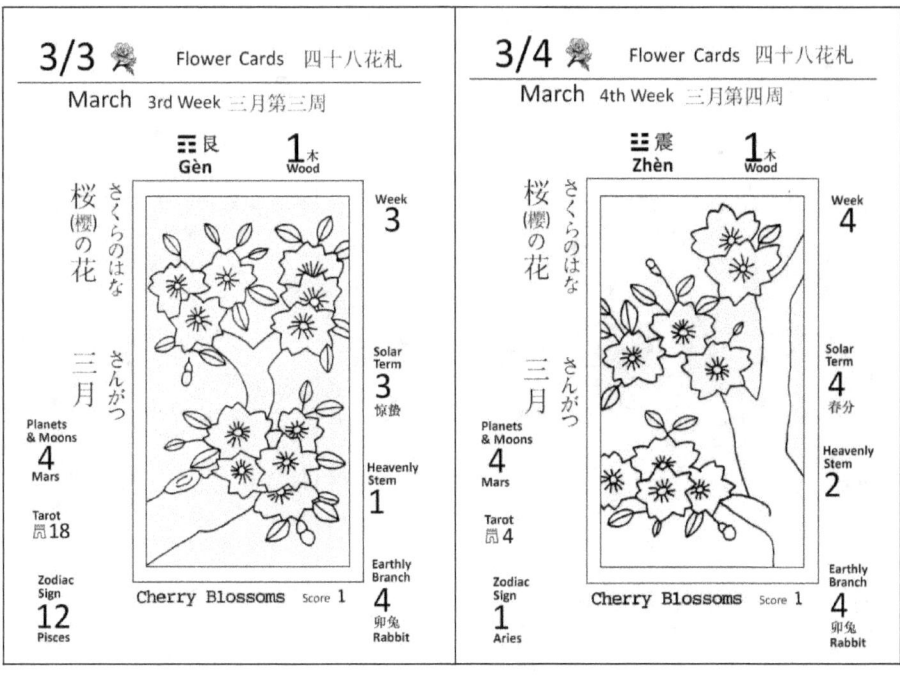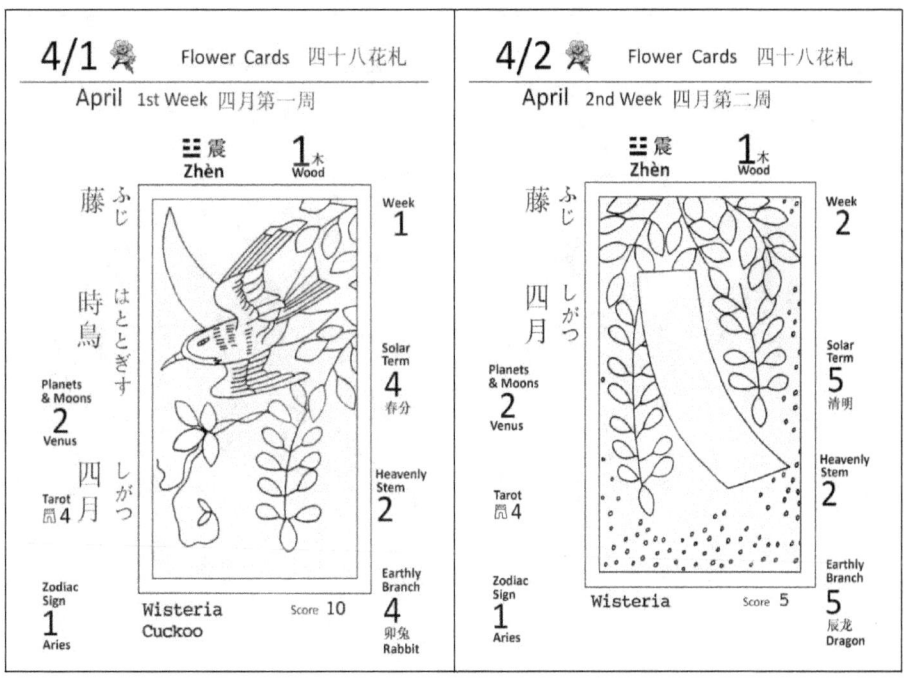

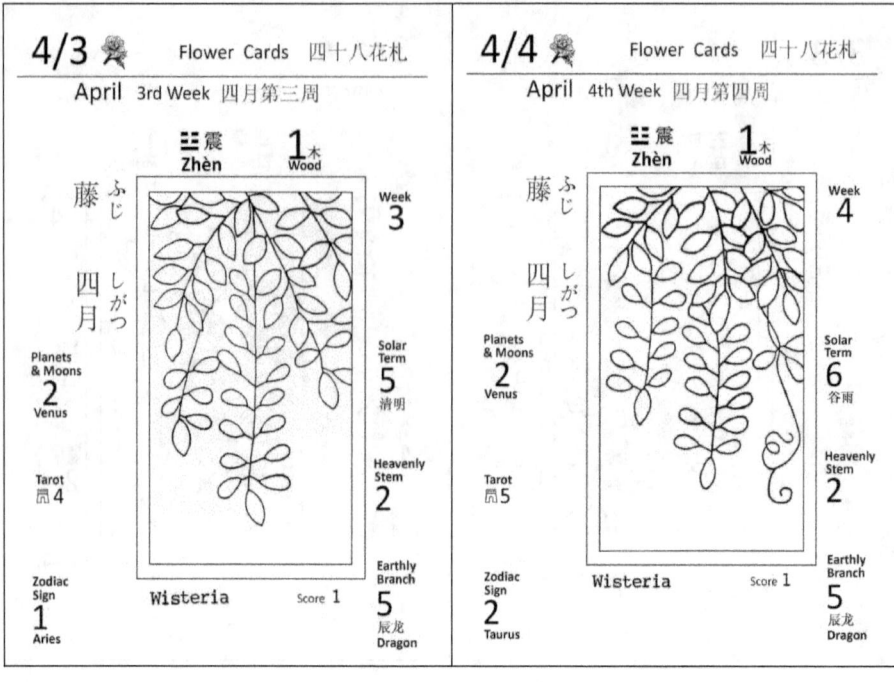

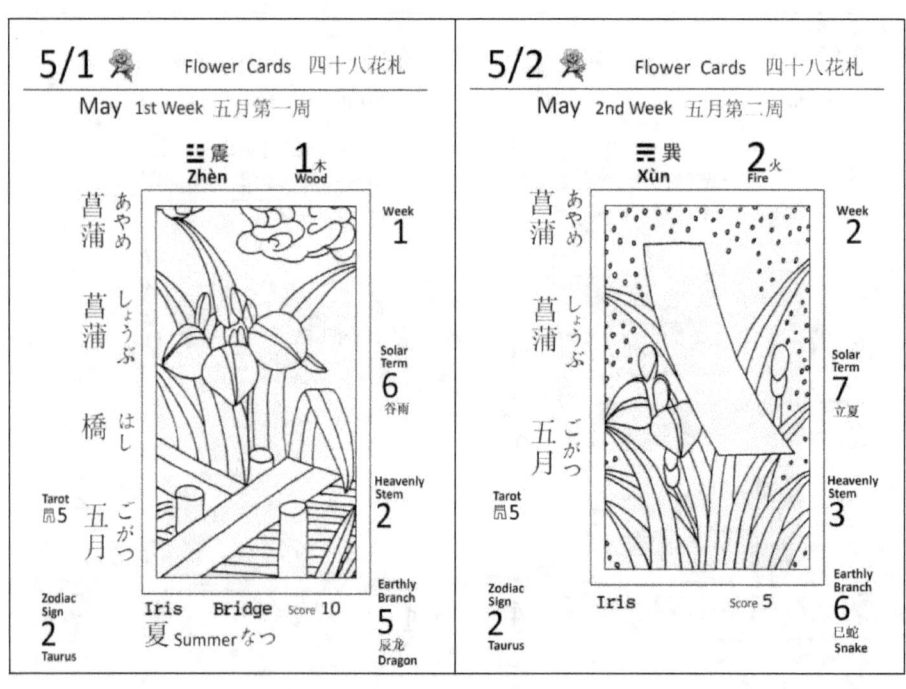

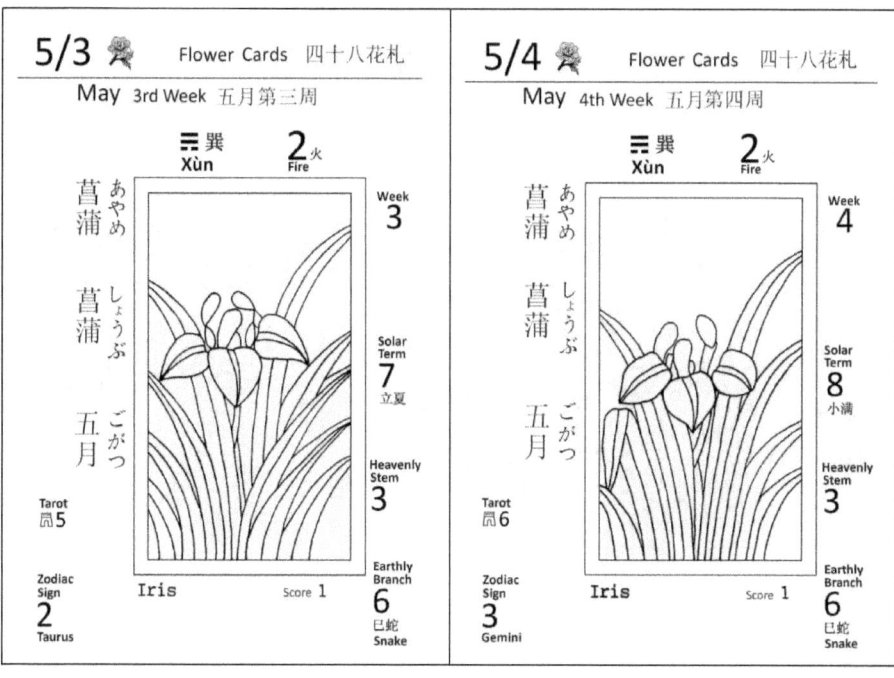

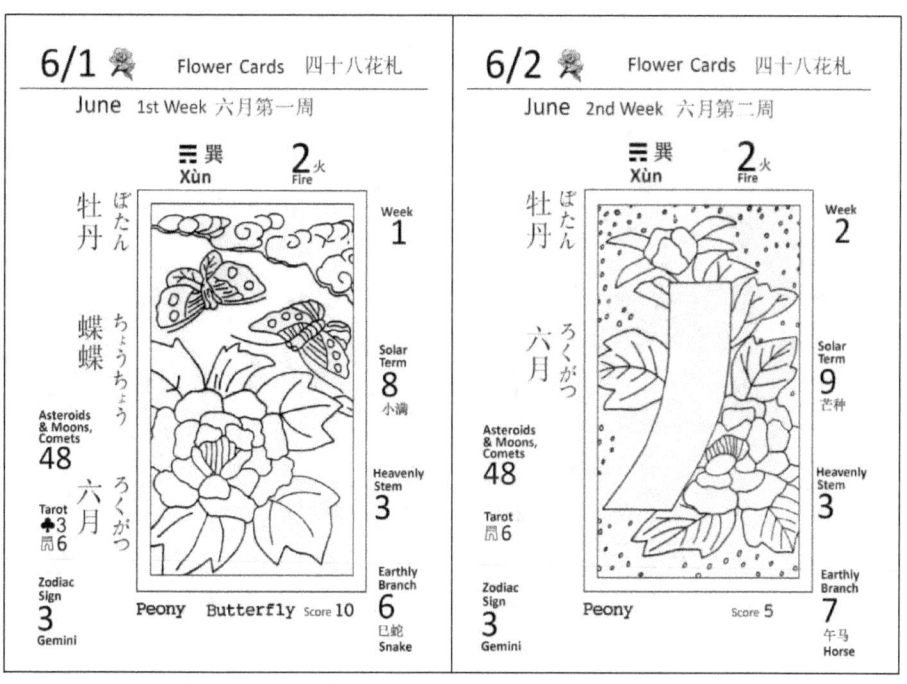

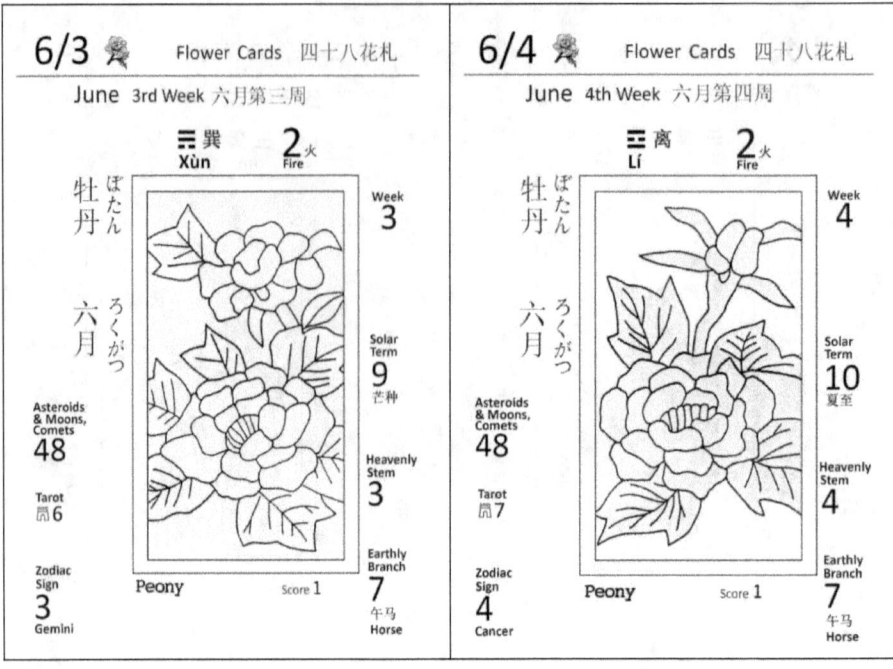

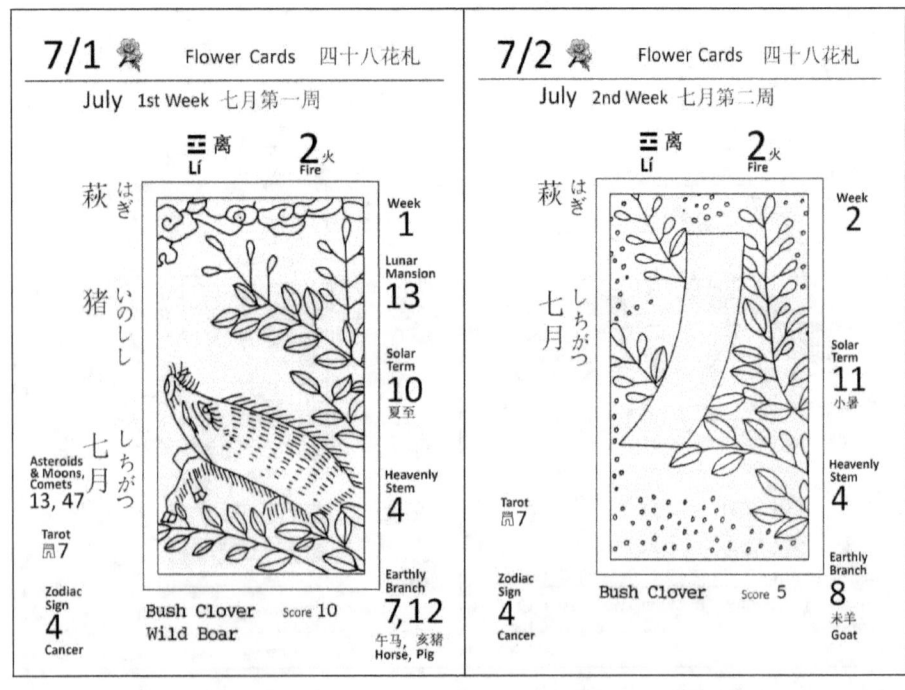

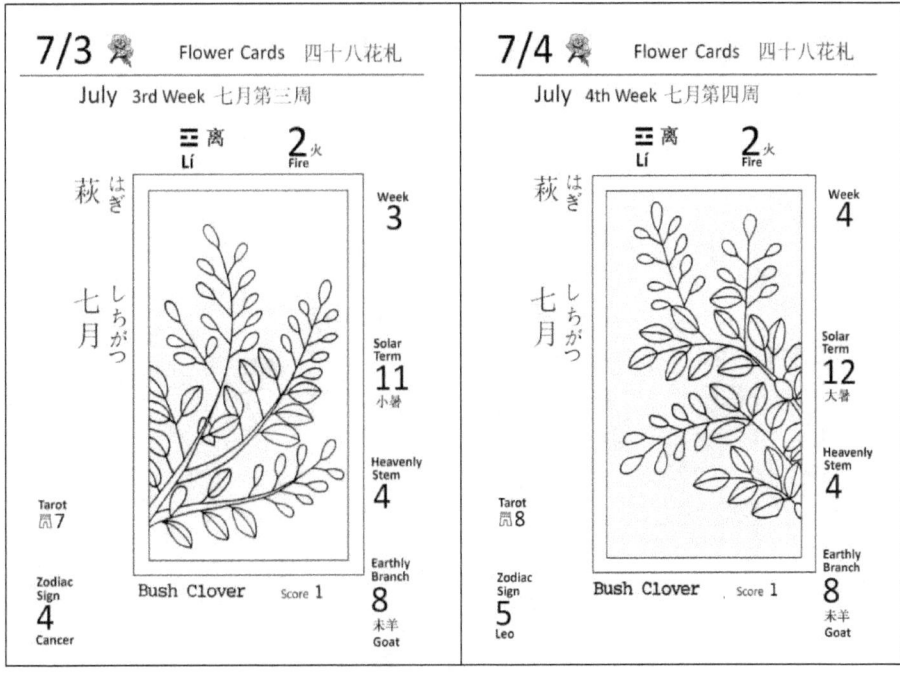

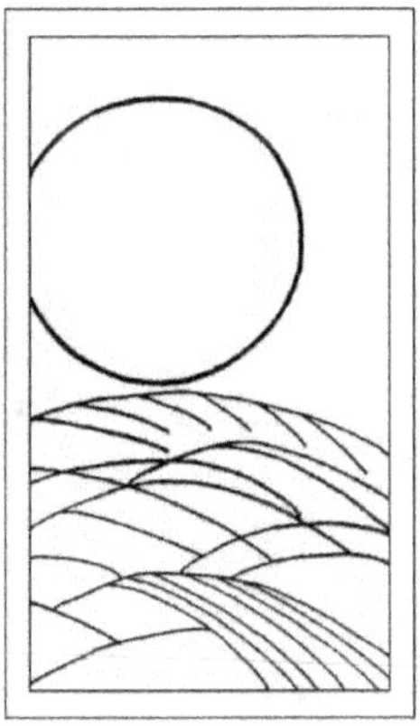

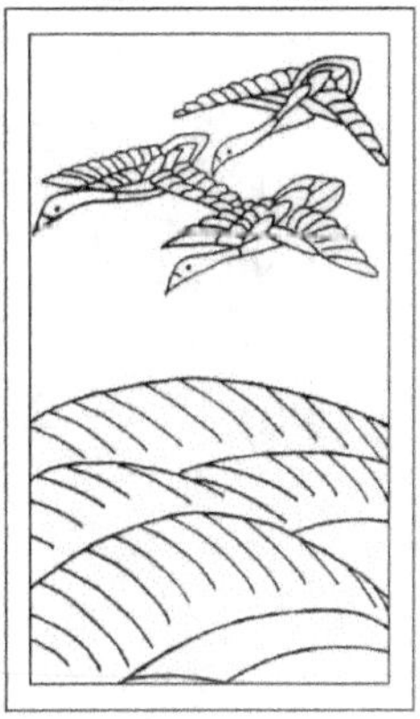

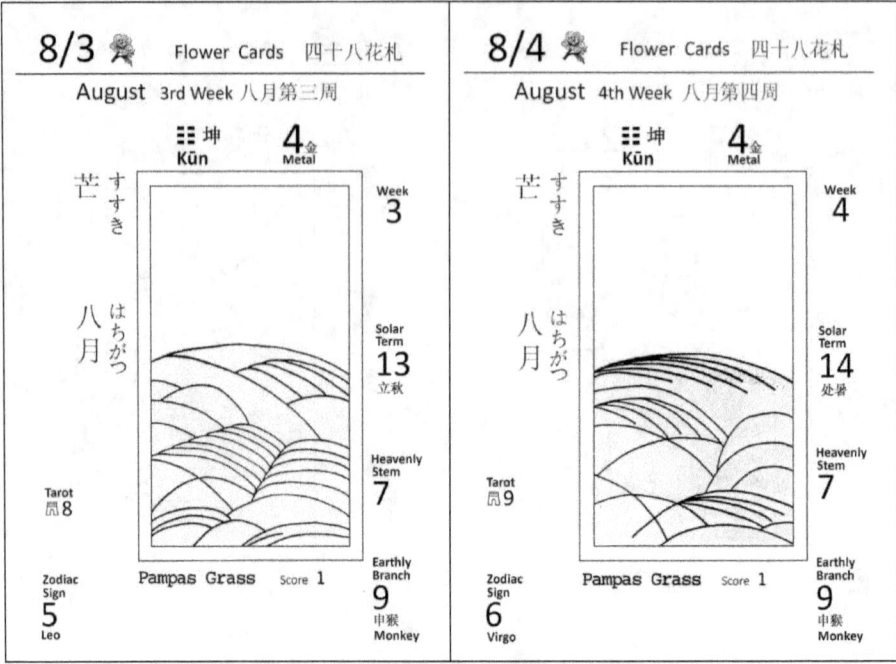
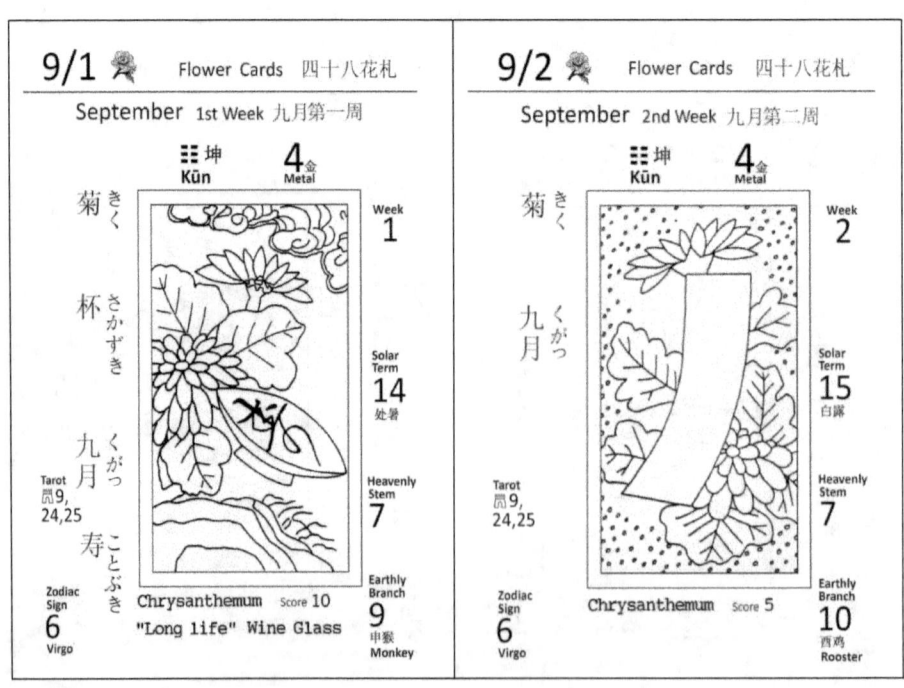

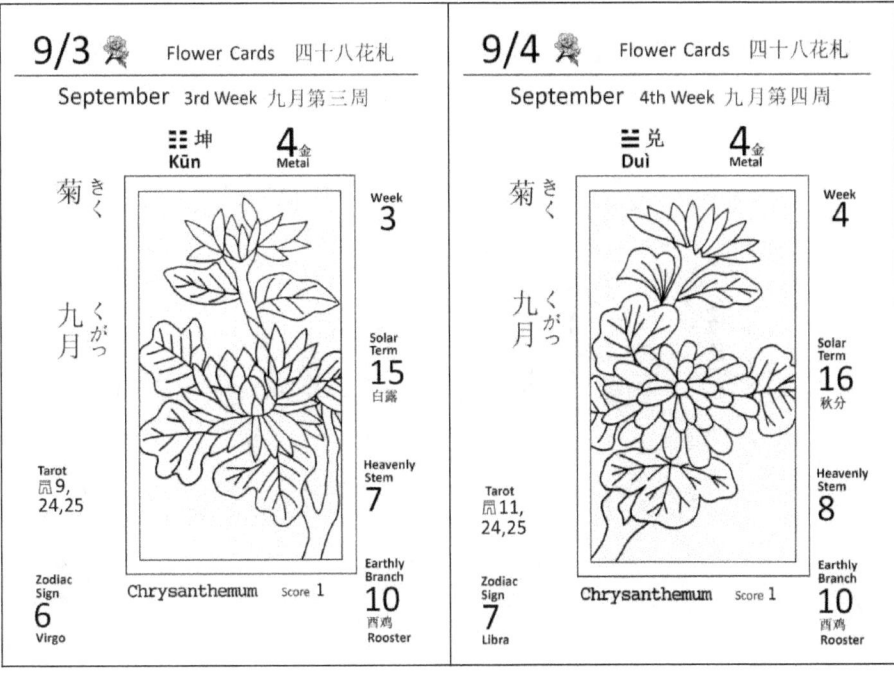

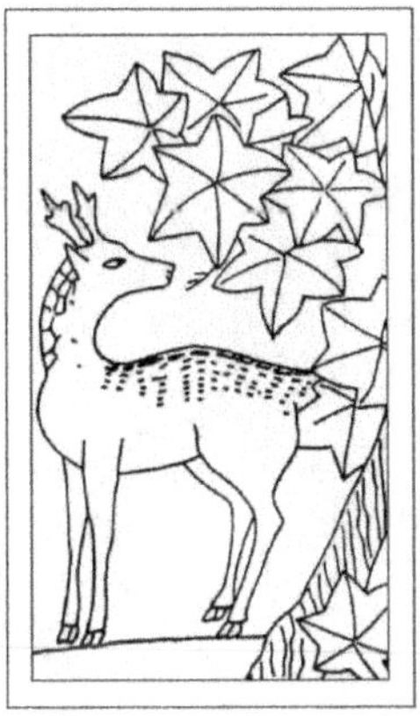

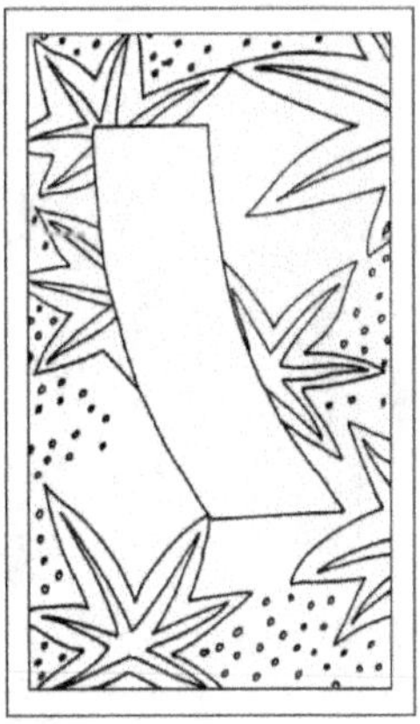

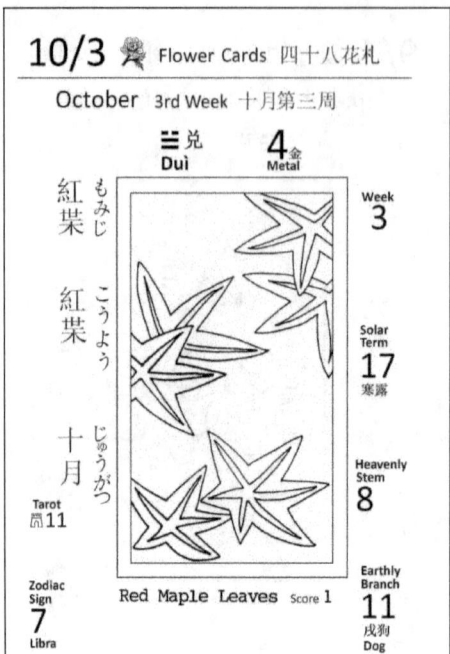
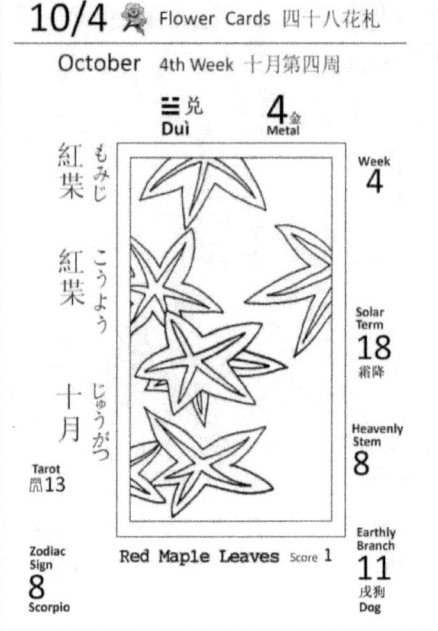

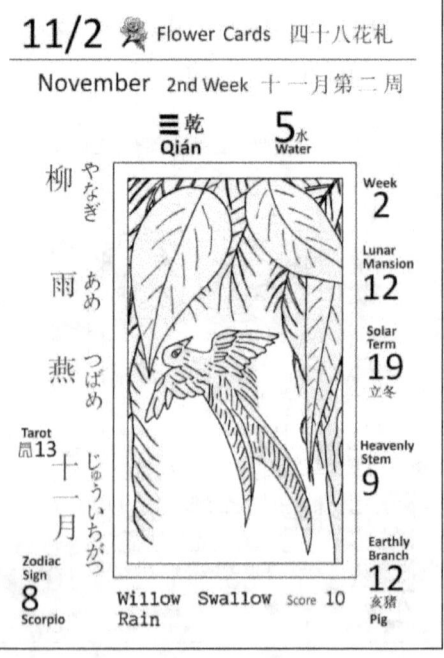

11/3 🌹 Flower Cards 四十八花札
November 3rd Week 十一月第三周

☰ 乾 Qián
5 水 Water

柳 やなぎ
雨 あめ
十一月 じゅういちがつ

Tarot 🂭 13

Zodiac Sign 8 Scorpio

Week 3
Solar Term 19 立冬
Heavenly Stem 9
Earthly Branch 12 亥猪 Pig

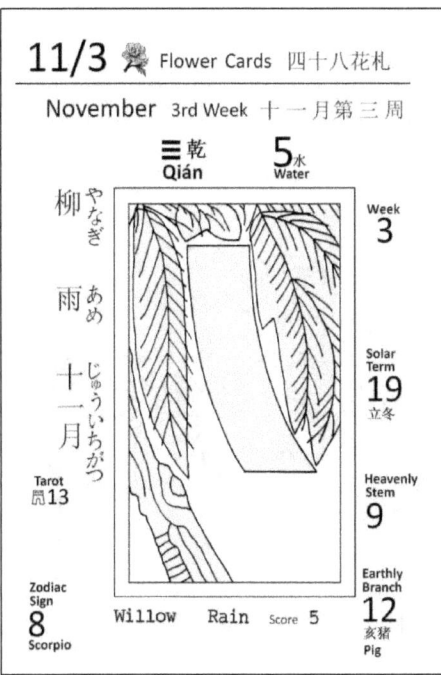

Willow Rain   Score 5

11/4 🌹 Flower Cards 四十八花札
November 4th Week 十一月第四周

Chemical Elements 90
Planets & Moons 6/12(2) 7/50

☰ 乾 Qián
5 水 Water

Tarot 🂭 14

柳 やなぎ
雨 あめ
雷 かみなり
電光 でんこう
十一月 じゅういちがつ

Zhèn ☳ Thunder

Zodiac Sign 9 Sagittarius

Week 4
Solar Term 20 小雪
Heavenly Stem 9
Earthly Branch 12 亥猪 Pig

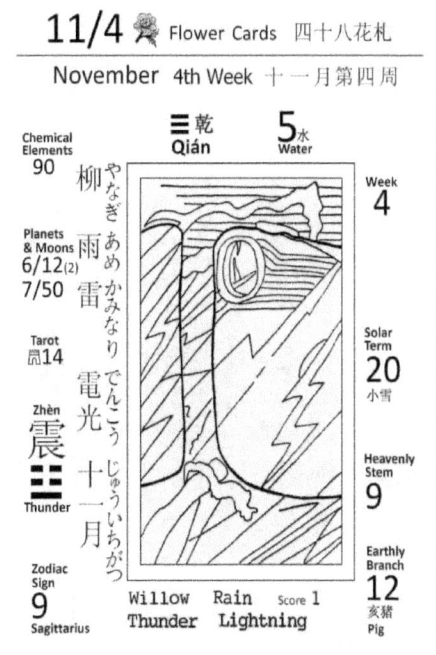

Willow Rain   Score 1
Thunder Lightning

12/1 🌹 Flower Cards 四十八花札
December 1st Week 十二月第一周

☰ 乾 Qián
5 水 Water

桐 きり

Conste-llation 4

鳳凰 ほうおう
十二月 じゅうにがつ

Tarot ♦27 🂭14 ♥10

Zodiac Sign 9 Sagittarius

Week 1
Solar Term 20 小雪
Heavenly Stem 9
Earthly Branch 12 亥猪 Pig

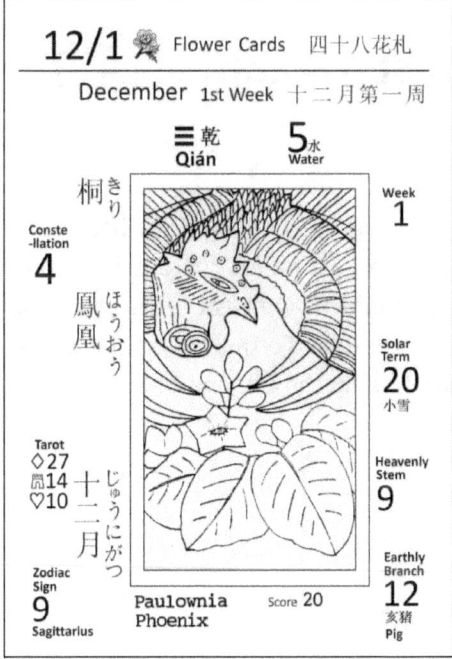

Paulownia   Score 20
Phoenix

12/2 🌹 Flower Cards 四十八花札
December 2nd Week 十二月第二周

☰ 乾 Qián
5 水 Water

桐 きり
十二月 じゅうにがつ

Tarot 🂭 14

Zodiac Sign 9 Sagittarius

Week 2
Solar Term 21 大雪
Heavenly Stem 9
Earthly Branch 1 子鼠 Rat

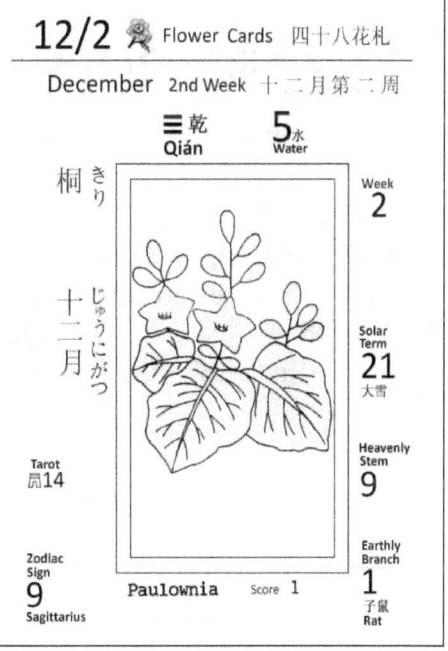

Paulownia   Score 1

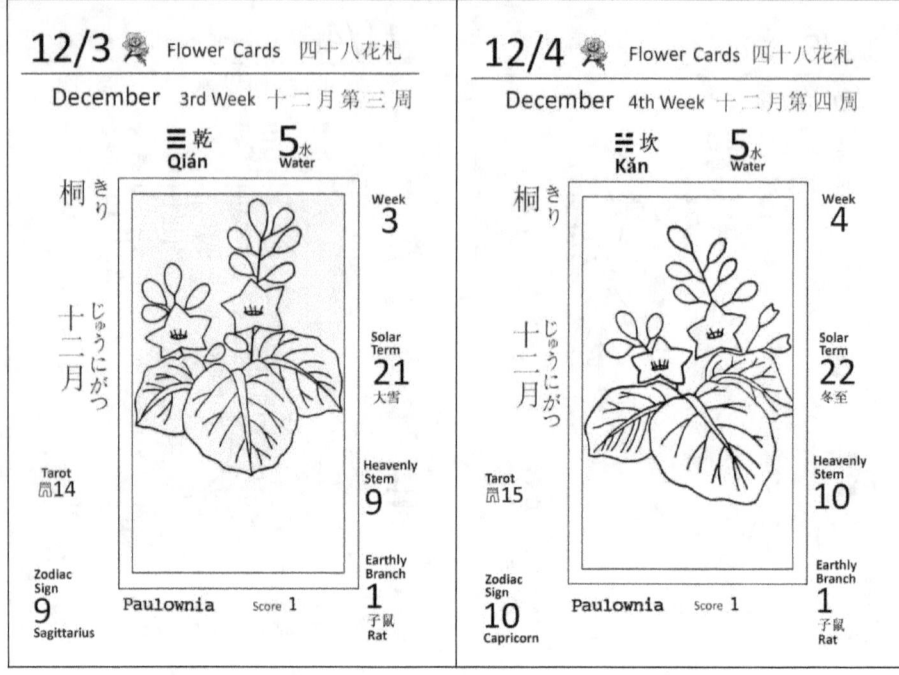

The 12 Signs of the Western Zodiac 西方十二星座

The Western Zodiac (西方十二星座 Xī Fāng Shí Èr Xīng Zuò) has signs which are relatable to the 28 Lunar Mansions, the 88 Constellations, the 12 Earthly Branches, the Asteroids and moons as well as the Tarot cards. Based on their close relationship to the 24 Solar Terms, the zodiac signs can be allotted a Chinese Element with 3 being the Change of Seasons.

The 10 Heavenly Stems 十天干

The Ten Heavenly Stems (十天干 Shí Tiān Gān) are a series of ordinal numbers which were originally used during the Shang Dynasty (商朝代 Shāng Cháo-dài, c. 1600 – 1046 BC) as names for the days of a ten-day week. They are therefore connected to the 10 suns with sun number 3 丙 Bǐng (Yáng Fire 阳火) being the one that survived the shooting by Hòu Yì 后羿. The Heavenly Stems are related to the Five Elements and Five Directions with every two stems connected to one element and one direction. There are also Tarot card connections.

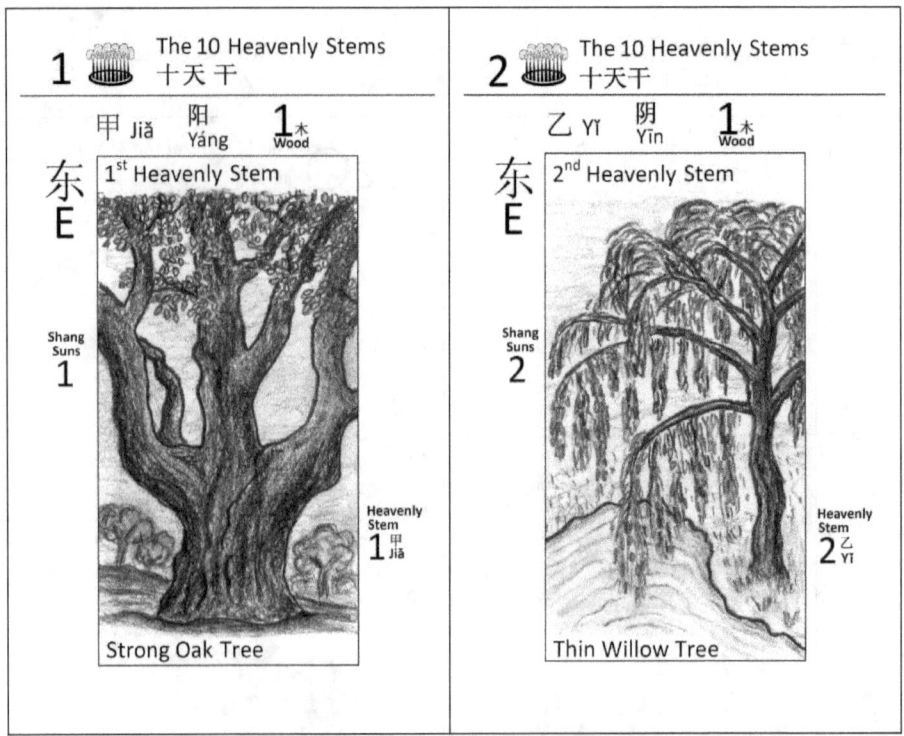

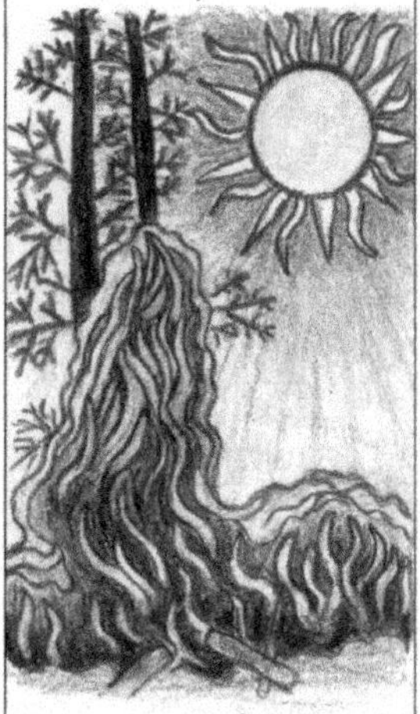

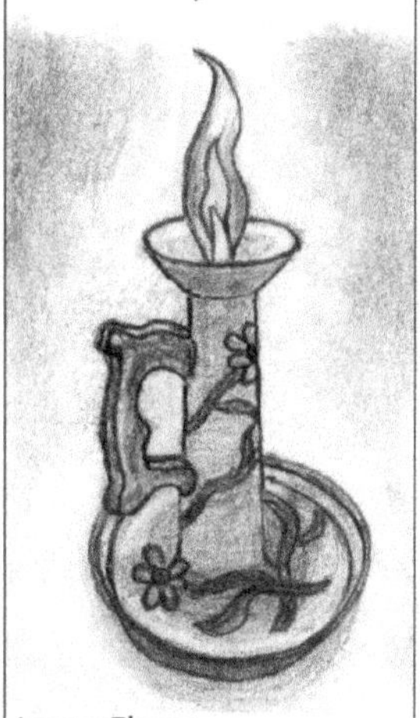

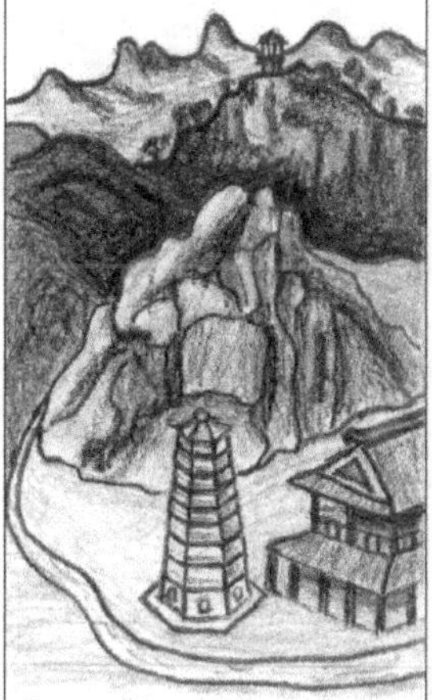

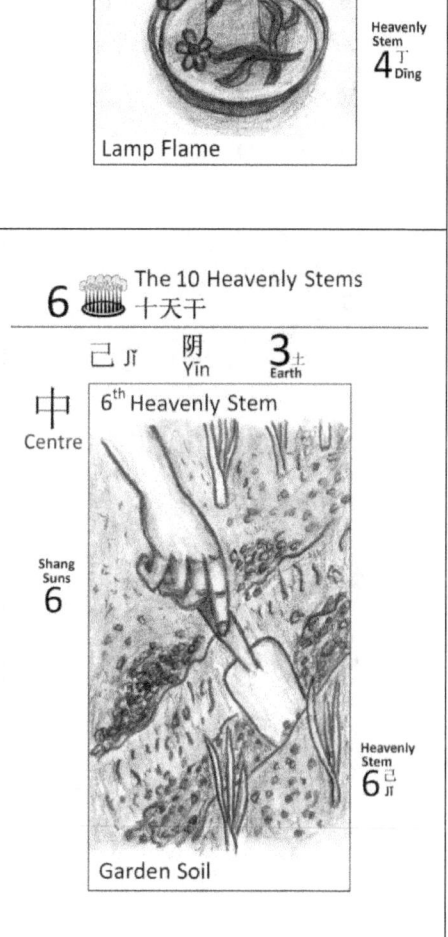

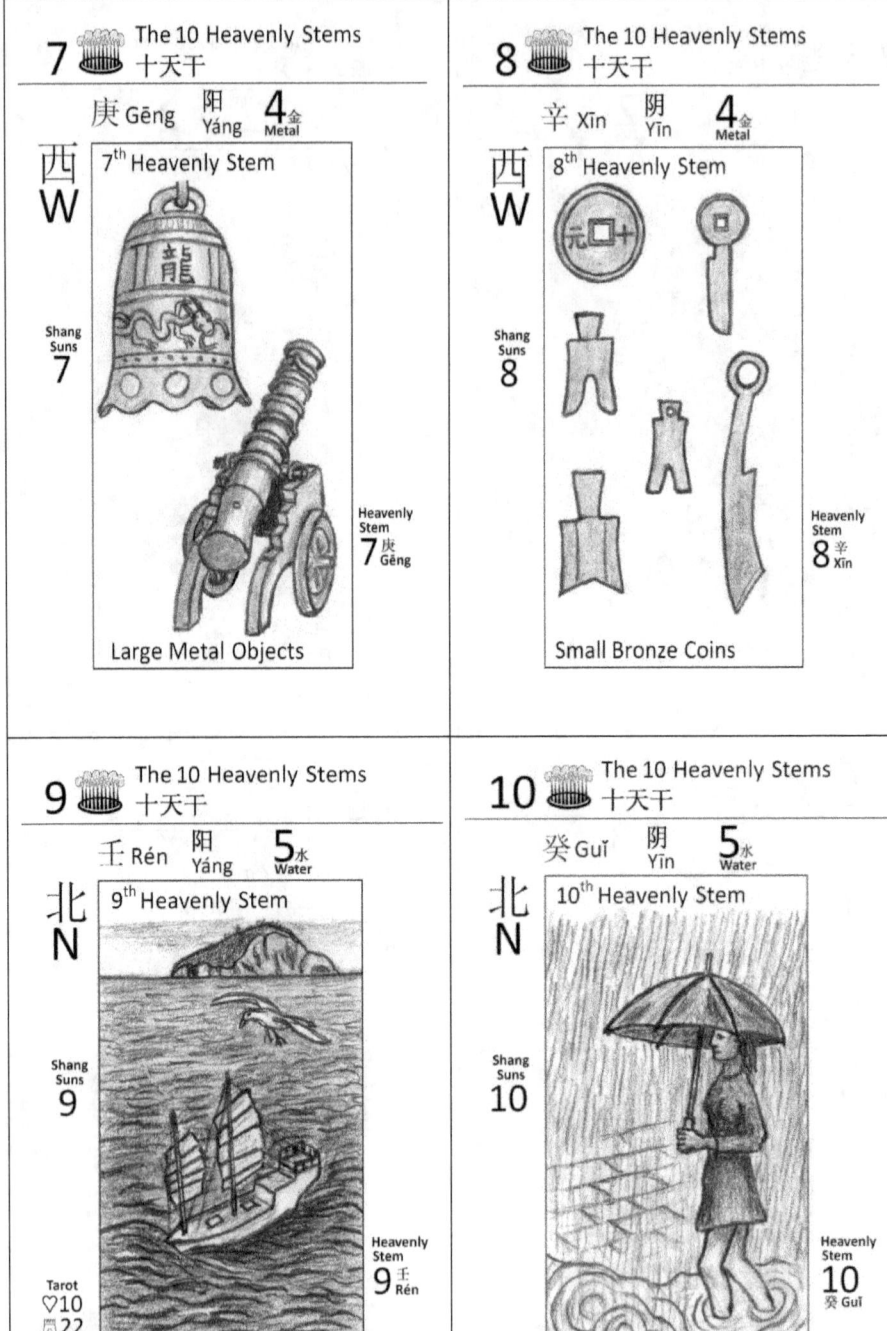

The 12 Earthly Branches 十二地支 (Shí Èr Dì Zhī) and the 12 Zodiacal Animals known in Chinese as 十二生肖 Shí Èr Shēng Xiào meaning "the 12 Birth Emblems" or 十二属相 (Shí Èr Shǔ Xiàng) meaning "the 12 Signs of Belonging"

There are connections to the 28 Lunar Mansions, the 88 Constellations, the Planets and Moons, the Asteroids and Moons and the Tarot cards.

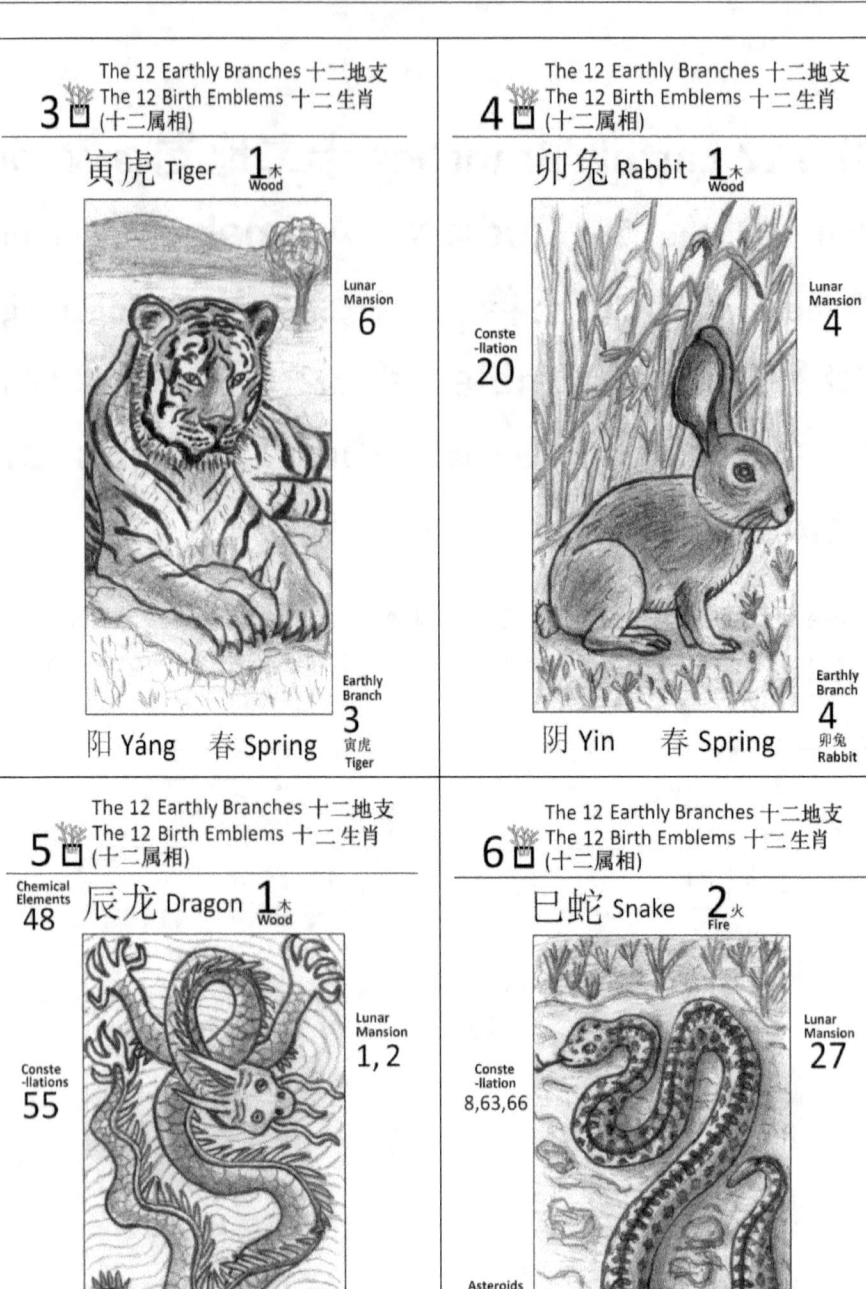

7 | The 12 Earthly Branches 十二地支
The 12 Birth Emblems 十二生肖
(十二属相)

Chemical Elements 12,25, 31,71

午马 Horse **2** 火 Fire

Lunar Mansion 25

Constellations 80,86

Planets & Moons 6/39, 7/20, 9/3

Asteroids & Moons, Comets 98, 111

Tarot
♠12
♣12
♢12
♡12

阳 Yáng 夏 Summer **Earthly Branch 7** 午马 Horse

8 | The 12 Earthly Branches 十二地支
The 12 Birth Emblems 十二生肖
(十二属相)

未羊 Goat **2** 火 Fire

Lunar Mansion 23

Constellation 24,79

Planets & Moons 6/2, 6/3, 6/24, 6/25

Zodiac Sign 10 Capricorn

阴 Yin 夏 Summer **Earthly Branch 8** 未羊 Goat

9 | The 12 Earthly Branches 十二地支
The 12 Birth Emblems 十二生肖
(十二属相)

申猴 Monkey **4** 金 Metal

Lunar Mansion 20

阳 Yáng 秋 Autumn **Earthly Branch 9** 申猴 Monkey

10 | The 12 Earthly Branches 十二地支
The 12 Birth Emblems 十二生肖
(十二属相)

Chemical Elements 87

酉鸡 Rooster **4** 金 Metal

Lunar Mansion 18

Tarot 🀫28

阴 Yin 秋 Autumn **Earthly Branch 10** 酉鸡 Rooster

The 60 Year Cycle 六十甲子

This 60 year cycle (六十甲子 Liù Shí Jiǎ Zǐ) begins on 8th February 1864 and ends on 4th February 1924 covering an interesting historical period without being involved in contentious modern politics. Sun Yat-sen (Sūn Yì Xiān 孙逸仙), the first Interim President of the Republic of China, decided in 1912 to introduce a system of continuously numbered years starting from the first year of the Yellow Emperor Huáng Dì 皇帝 traditionally believed to be 2698 BC. Using this system the period covered is the 77th cycle from the year 4562 to 4621. The 60 years are of course related to the 10 Heavenly Stems and the 12 Earthly Branches and also to Yin and Yang, the 5 Elements and the Seven Northern Dipper Stars 北斗七星.

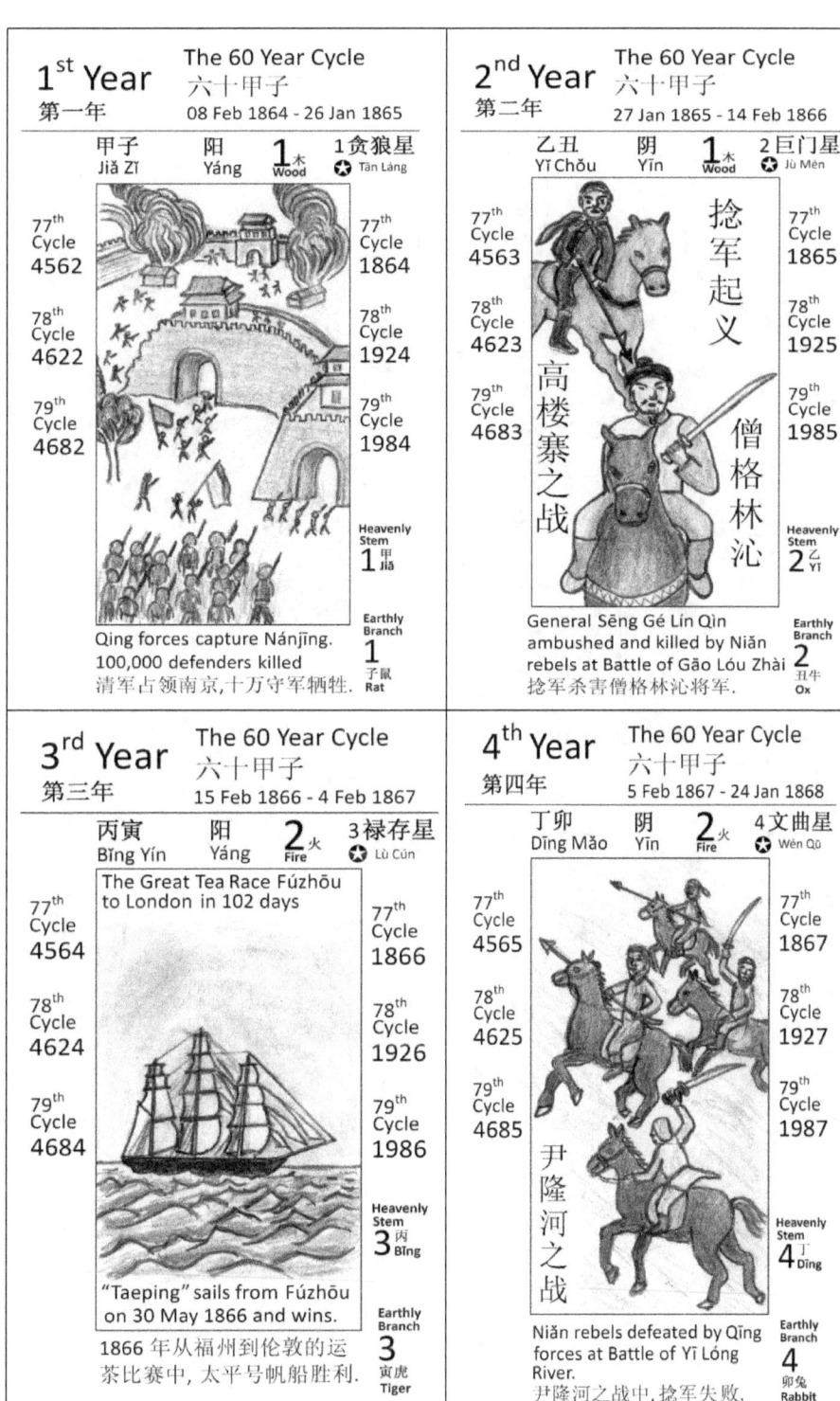

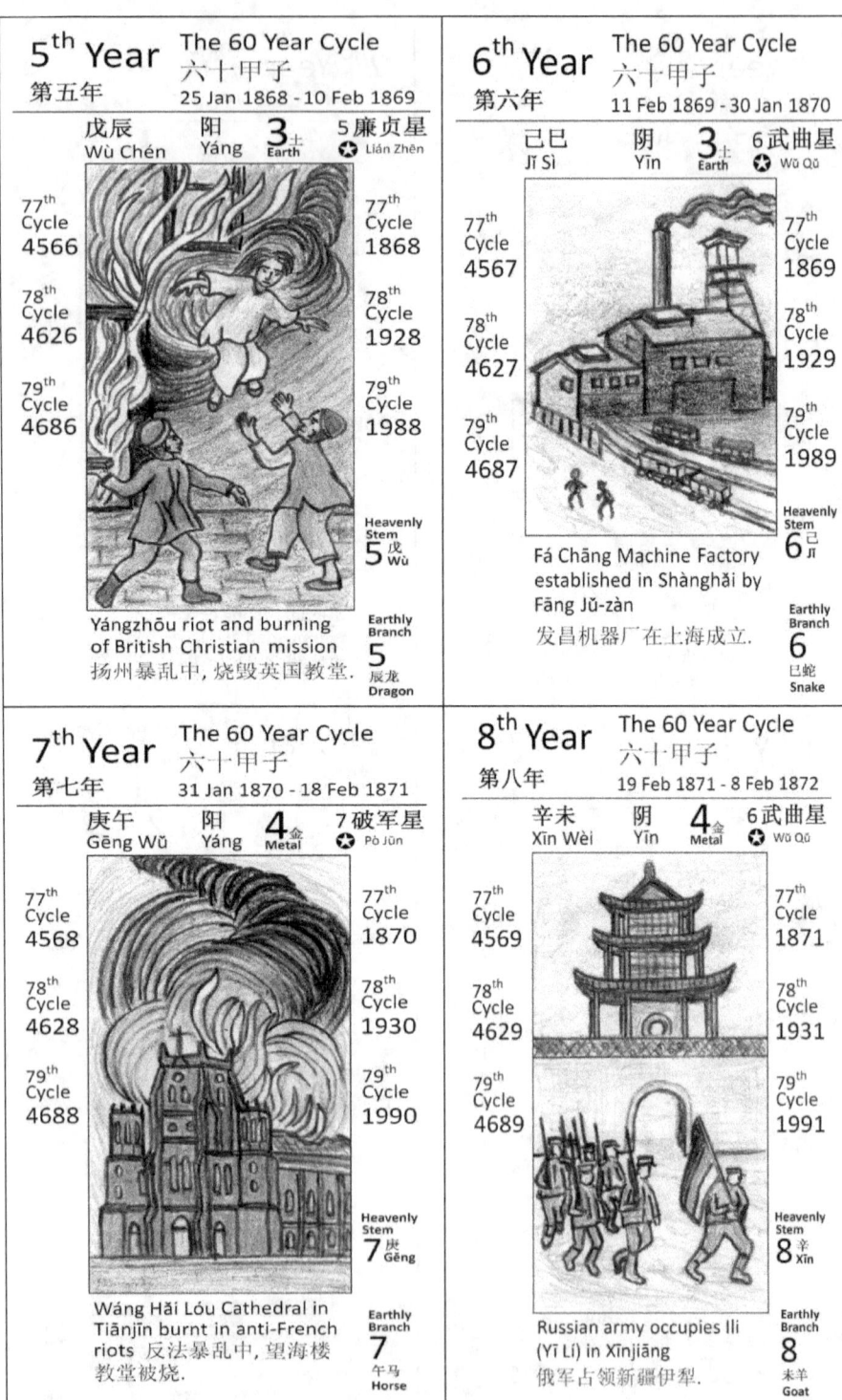

9th Year 第九年
The 60 Year Cycle 六十甲子
9 Feb 1872 - 28 Jan 1873

壬申 Rén Shēn | 阳 Yáng | 5 Water 水 | ★ 5 廉贞星 Lián Zhēn

容闳 1828-1912

77th Cycle 4570	77th Cycle 1872
78th Cycle 4630	78th Cycle 1932
79th Cycle 4690	79th Cycle 1992

Yung Wing (Róng Hóng) founds Chinese Educational Mission and sends students to America 容闳率学生赴美留学, 史称"中国幼童留美运动".

Heavenly Stem **9** 壬 Rén
Earthly Branch **9** 申猴 Monkey

10th Year 第十年
The 60 Year Cycle 六十甲子
29 Jan 1873 - 16 Feb 1874

癸酉 Guǐ Yǒu | 阴 Yīn | 5 Water 水 | ★ 4 文曲星 Wén Qǔ

77th Cycle 4571	77th Cycle 1873
78th Cycle 4631	78th Cycle 1933
79th Cycle 4691	79th Cycle 1993

China Merchant Steamship Navigation Company opens in Shànghǎi. 中国轮船招商局在上海设立.

Heavenly Stem **10** 癸 Guǐ
Earthly Branch **10** 酉鸡 Rooster

11th Year 第十一年
The 60 Year Cycle 六十甲子
17 Feb 1874 - 5 Feb 1875

甲戌 Jiǎ Xū | 阳 Yáng | 1 Wood 木 | ★ 3 禄存星 Lù Cún

77th Cycle 4572	77th Cycle 1874
78th Cycle 4632	78th Cycle 1934
79th Cycle 4692	79th Cycle 1994

May-Dec 1874, Japanese flagship Ryūjō and 3,600 soldiers engage in punitive attack on aborigines in Táiwān.
1874 年 5 月至 12 月, 牡丹社事件中龙骧号旗舰船及 3600 名日军报复性攻打台湾原住民.

Heavenly Stem **1** 甲 Jiǎ
Earthly Branch **11** 戌狗 Dog

12th Year 第十二年
The 60 Year Cycle 六十甲子
6 Feb 1875 - 25 Jan 1876

乙亥 Yǐ Hài | 阴 Yīn | 1 Wood 木 | ★ 2 巨门星 Jù Mén

77th Cycle 4573	77th Cycle 1875
78th Cycle 4633	78th Cycle 1935
79th Cycle 4693	79th Cycle 1995

Zuǒ Zōng-táng begins reconquest of Xīnjiāng. 左宗堂开始收复新疆.

Heavenly Stem **2** 乙 Yǐ
Earthly Branch **12** 亥猪 Pig

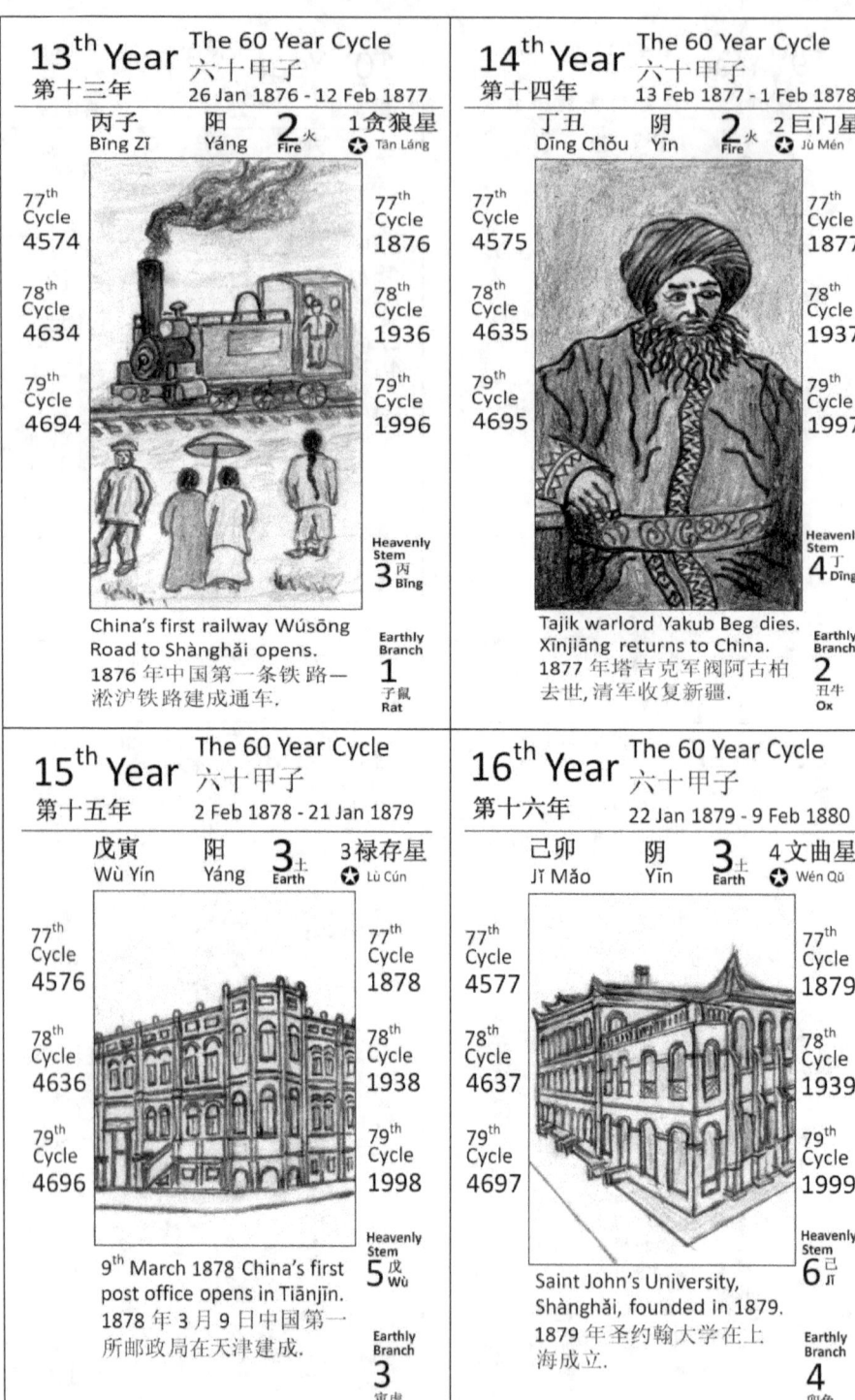

13th Year 第十三年
The 60 Year Cycle 六十甲子
26 Jan 1876 - 12 Feb 1877

丙子 Bǐng Zǐ | 阳 Yáng | 2 Fire | 1 贪狼星 Tān Láng

- 77th Cycle 4574
- 78th Cycle 4634
- 79th Cycle 4694

- 77th Cycle 1876
- 78th Cycle 1936
- 79th Cycle 1996

Heavenly Stem 3 丙 Bǐng
Earthly Branch 1 子鼠 Rat

China's first railway Wúsōng Road to Shànghǎi opens.
1876 年中国第一条铁路—淞沪铁路建成通车.

14th Year 第十四年
The 60 Year Cycle 六十甲子
13 Feb 1877 - 1 Feb 1878

丁丑 Dīng Chǒu | 阴 Yīn | 2 Fire | 2 巨门星 Jù Mén

- 77th Cycle 4575
- 78th Cycle 4635
- 79th Cycle 4695

- 77th Cycle 1877
- 78th Cycle 1937
- 79th Cycle 1997

Heavenly Stem 4 丁 Dīng
Earthly Branch 2 丑牛 Ox

Tajik warlord Yakub Beg dies. Xīnjiāng returns to China.
1877 年塔吉克军阀阿古柏去世,清军收复新疆.

15th Year 第十五年
The 60 Year Cycle 六十甲子
2 Feb 1878 - 21 Jan 1879

戊寅 Wù Yín | 阳 Yáng | 3 Earth | 3 禄存星 Lù Cún

- 77th Cycle 4576
- 78th Cycle 4636
- 79th Cycle 4696

- 77th Cycle 1878
- 78th Cycle 1938
- 79th Cycle 1998

Heavenly Stem 5 戊 Wù
Earthly Branch 3 寅虎 Tiger

9th March 1878 China's first post office opens in Tiānjīn.
1878 年 3 月 9 日中国第一所邮政局在天津建成.

16th Year 第十六年
The 60 Year Cycle 六十甲子
22 Jan 1879 - 9 Feb 1880

己卯 Jǐ Mǎo | 阴 Yīn | 3 Earth | 4 文曲星 Wén Qū

- 77th Cycle 4577
- 78th Cycle 4637
- 79th Cycle 4697

- 77th Cycle 1879
- 78th Cycle 1939
- 79th Cycle 1999

Heavenly Stem 6 己 Jǐ
Earthly Branch 4 卯兔 Rabbit

Saint John's University, Shànghǎi, founded in 1879.
1879 年圣约翰大学在上海成立.

17th Year 第十七年

The 60 Year Cycle 六十甲子
10 Feb 1880 - 29 Jan 1881

庚辰 Gēng Chén | 阳 Yáng | 4 金 Metal | 5 廉贞星 Lián Zhēn

77th Cycle 4578
78th Cycle 4638
79th Cycle 4698

77th Cycle 1880
78th Cycle 1940
79th Cycle 2000

Chinese cruiser Chāo Yǒng launched 11 Nov 1880.
中国超勇号舰于 1880 年 11 月 11 日下水.

Heavenly Stem **7** 庚 Gēng

Earthly Branch **5** 辰龙 Dragon

18th Year 第十八年

The 60 Year Cycle 六十甲子
30 Jan 1881 - 17 Feb 1882

辛巳 Xīn Sì | 阴 Yīn | 4 金 Metal | 6 武曲星 Wǔ Qū

77th Cycle 4579
78th Cycle 4639
79th Cycle 4699

77th Cycle 1881
78th Cycle 1941
79th Cycle 2001

First steam locomotive made in China entered service on 9th June 1881.
1881 年 6 月 9 日第一台中国制造的蒸汽机车投入运行.

Heavenly Stem **8** 辛 Xīn

Earthly Branch **6** 巳蛇 Snake

19th Year 第十九年

The 60 Year Cycle 六十甲子
18 Feb 1882 - 7 Feb 1883

壬午 Rén Wǔ | 阳 Yáng | 5 水 Water | 7 破军星 Pò Jūn

77th Cycle 4580
78th Cycle 4640
79th Cycle 4700

77th Cycle 1882
78th Cycle 1942
79th Cycle 2002

Jade Buddha Temple in Shànghǎi completed 1882.
1882 年上海玉佛寺建成.

Heavenly Stem **9** 壬 Rén

Earthly Branch **7** 午马 Horse

20th Year 第二十年

The 60 Year Cycle 六十甲子
8 Feb 1883 - 27 Jan 1884

癸未 Guǐ Wèi | 阴 Yīn | 5 水 Water | 6 武曲星 Wǔ Qū

77th Cycle 4581
78th Cycle 4641
79th Cycle 4701

77th Cycle 1883
78th Cycle 1943
79th Cycle 2003

16 Dec 1883 French troops capture Son Tay (Shān Xī Zhèn) from China. 1883 年12月16日 法军占领中国的山西镇.

Heavenly Stem **10** 癸 Guǐ

Earthly Branch **8** 未羊 Goat

21st Year 第二十一年
The 60 Year Cycle 六十甲子
28 Jan 1884 - 14 Feb 1885

甲申 Jiǎ Shēn | 阳 Yáng | 1 Wood | 5 廉贞星 Lián Zhēn

77th Cycle 4582 | 77th Cycle 1884
78th Cycle 4642 | 78th Cycle 1944
79th Cycle 4702 | 79th Cycle 2004

Heavenly Stem 1 甲 Jiǎ
Earthly Branch 9 申猴 Monkey

23 June 1884 French retreat from Chinese ambush at Bac Le (Běi Lí). 1884 年 6 月 23 日清军在北梨伏击法军,致其撤退.

22nd Year 第二十二年
The 60 Year Cycle 六十甲子
15 Feb 1885 - 3 Feb 1886

乙酉 Yǐ Yǒu | 阴 Yīn | 1 Wood | 4 文曲星 Wén Qū

镇南关之战

77th Cycle 4583 | 77th Cycle 1885
78th Cycle 4643 | 78th Cycle 1945
79th Cycle 4703 | 79th Cycle 2005

Sino-French War 中法战争

Heavenly Stem 2 乙 Yǐ
Earthly Branch 10 酉鸡 Rooster

Chinese victory at Battle of Zhèn Nán Pass 23 March 1885. 1885 年 3 月 23 日镇南关之战, 中方胜利.

23rd Year 第二十三年
The 60 Year Cycle 六十甲子
4 Feb 1886 - 23 Jan 1887

丙戌 Bǐng Xū | 阳 Yáng | 2 Fire | 3 禄存星 Lù Cún

77th Cycle 4584 | 77th Cycle 1886
78th Cycle 4644 | 78th Cycle 1946
79th Cycle 4704 | 79th Cycle 2006

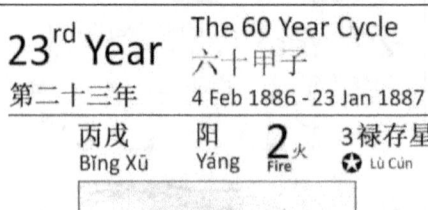

Chinese cruiser Jìng Yuǎn launched 14 Dec 1886. 中国靖远号舰于 1886 年 12 月 14 日下水.

Heavenly Stem 3 丙 Bǐng
Earthly Branch 11 戌狗 Dog

24th Year 第二十四年
The 60 Year Cycle 六十甲子
24 Jan 1887 - 11 Feb 1888

丁亥 Dīng Hài | 阴 Yīn | 2 Fire | 2 巨门星 Jù Mén

77th Cycle 4585 | 77th Cycle 1887
78th Cycle 4645 | 78th Cycle 1947
79th Cycle 4705 | 79th Cycle 2007

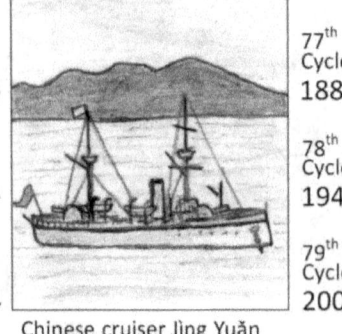

Hong Kong College of Medicine for Chinese established 1887. 1887 年香港华人西医书院成立.

Heavenly Stem 4 丁 Dīng
Earthly Branch 12 亥猪 Pig

25th Year 第二十五年
The 60 Year Cycle 六十甲子
12 Feb 1888 - 30 Jan 1889

戊子 Wù Zǐ | 阳 Yáng | **3** 土 Earth | 1 贪狼星 Tān Láng

77th Cycle 4586
78th Cycle 4646
79th Cycle 4706

77th Cycle 1888
78th Cycle 1948
79th Cycle 2008

Heavenly Stem **5** 戊 Wù
Earthly Branch **1** 子鼠 Rat

Star Ferry Company founded in Hong Kong in 1888.
1888 年香港天星小轮公司成立.

26th Year 第二十六年
The 60 Year Cycle 六十甲子
31 Jan 1889 - 20 Jan 1890

己丑 Jǐ Chǒu | 阴 Yīn | **3** 土 Earth | 2 巨门星 Jù Mén

77th Cycle 4587
78th Cycle 4647
79th Cycle 4707

77th Cycle 1889
78th Cycle 1949
79th Cycle 2009

香港高尔夫球会

Heavenly Stem **6** 己 Jǐ
Earthly Branch **2** 丑牛 Ox

Hong Kong Golf Club founded in 1889. 1889 年香港高尔夫球会成立.

27th Year 第二十七年
The 60 Year Cycle 六十甲子
21 Jan 1890 - 8 Feb 1891

庚寅 Gēng Yín | 阳 Yáng | **4** 金 Metal | 3 禄存星 Lù Cún

77th Cycle 4588
78th Cycle 4648
79th Cycle 4708

77th Cycle 1890
78th Cycle 1950
79th Cycle 2010

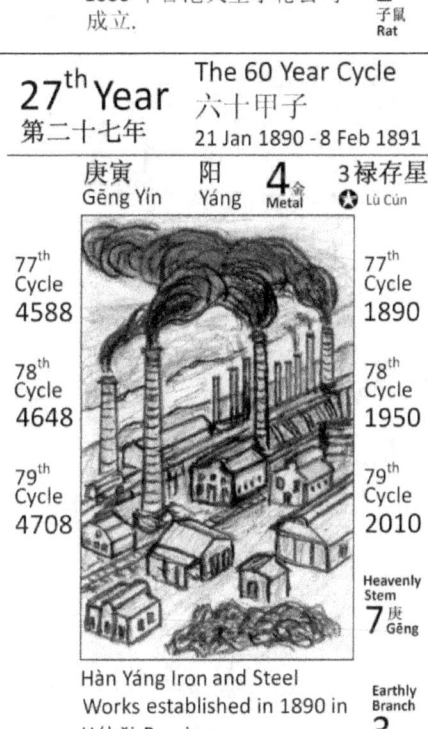

Heavenly Stem **7** 庚 Gēng
Earthly Branch **3** 寅虎 Tiger

Hàn Yáng Iron and Steel Works established in 1890 in Húběi Province.
1890 年湖北汉阳铁厂成立.

28th Year 第二十八年
The 60 Year Cycle 六十甲子
9 Feb 1891 - 29 Jan 1892

辛卯 Xīn Mǎo | 阴 Yīn | **4** 金 Metal | 4 文曲星 Wén Qū

77th Cycle 4589
78th Cycle 4649
79th Cycle 4709

77th Cycle 1891
78th Cycle 1951
79th Cycle 2011

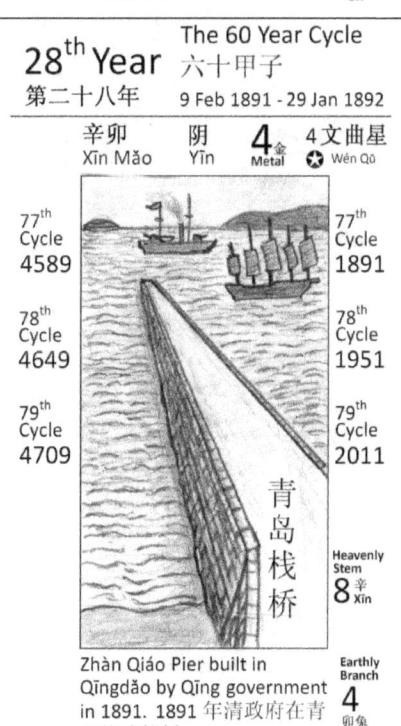

青岛栈桥

Heavenly Stem **8** 辛 Xīn
Earthly Branch **4** 卯兔 Rabbit

Zhàn Qiáo Pier built in Qīngdǎo by Qīng government in 1891. 1891 年清政府在青岛修建栈桥.

33rd Year 第三十三年
The 60 Year Cycle 六十甲子
13 Feb 1896 - 1 Feb 1897

丙申 Bǐng Shēn | 阳 Yáng | 2 火 Fire | 5 廉贞星 Lián Zhēn

- 77th Cycle 4594
- 78th Cycle 4654
- 79th Cycle 4714

- 77th Cycle 1896
- 78th Cycle 1956
- 79th Cycle 2016

孙中山伦敦蒙难

Sun Yat-sen kidnapped by Chinese Embassy in London. 1896 年孙中山在伦敦被中国使馆绑架.

Heavenly Stem 3 丙 Bǐng
Earthly Branch 9 申猴 Monkey

34th Year 第三十四年
The 60 Year Cycle 六十甲子
2 Feb 1897 - 21 Jan 1898

丁酉 Dīng Yǒu | 阴 Yīn | 2 火 Fire | 4 文曲星 Wén Qū

- 77th Cycle 4595
- 78th Cycle 4655
- 79th Cycle 4715

- 77th Cycle 1897
- 78th Cycle 1957
- 79th Cycle 2017

Germans land at Jiāo Zhōu Bay, Qīngdǎo 14 Nov 1897. 1897 年 11 月 14 日德国强占胶州湾.

Heavenly Stem 4 丁 Dīng
Earthly Branch 10 酉鸡 Rooster

35th Year 第三十五年
The 60 Year Cycle 六十甲子
22 Jan 1898 - 9 Feb 1899

戊戌 Wù Xū | 阳 Yáng | 3 土 Earth | 3 禄存星 Lù Cún

- 77th Cycle 4596
- 78th Cycle 4656
- 79th Cycle 4716

- 77th Cycle 1898
- 78th Cycle 1958
- 79th Cycle 2018

Russia starts building Chinese Eastern Railway in 1898. 1898 年俄国开始修建中东铁路.

Heavenly Stem 5 戊 Wù
Earthly Branch 11 戌狗 Dog

36th Year 第三十六年
The 60 Year Cycle 六十甲子
10 Feb 1899 - 30 Jan 1900

己亥 Jǐ Hài | 阴 Yīn | 3 土 Earth | 2 巨门星 Jù Mén

- 77th Cycle 4597
- 78th Cycle 4657
- 79th Cycle 4717

- 77th Cycle 1899
- 78th Cycle 1959
- 79th Cycle 2019

1899 年 10 月 18 日森罗殿之战清军击败义和团.

Boxers defeated by Qīng soldiers at Battle of Sēn Luó Temple on 18th Oct 1899.

Heavenly Stem 6 己 Jǐ
Earthly Branch 12 亥猪 Pig

37th Year 第三十七年

The 60 Year Cycle 六十甲子
31 Jan 1900 - 18 Feb 1901

庚子 Gēng Zǐ | 阳 Yáng | 4 金 Metal | 1 贪狼星 Tān Láng

- 77th Cycle 4598 / 1900
- 78th Cycle 4658 / 1960
- 79th Cycle 4718 / 2020

Heavenly Stem 7 庚 Gēng
Earthly Branch 1 子鼠 Rat

The Eight Powers burn Chéng Hǎi Lóu Temple at Shān-hǎi Guān, September 1900.
1900 年 9 月,八国联军火烧山海关澄海楼.

38th Year 第三十八年

The 60 Year Cycle 六十甲子
19 Feb 1901 - 7 Feb 1902

辛丑 Xīn Chǒu | 阴 Yīn | 4 金 Metal | 2 巨门星 Jù Mén

- 77th Cycle 4599 / 1901
- 78th Cycle 4659 / 1961
- 79th Cycle 4719 / 2021

Heavenly Stem 8 辛 Xīn
Earthly Branch 2 丑牛 Ox

The Eleven Powers and China sign the Boxer Protocol for settlement of the Boxer Uprising 7 September 1901.
1901 年 9 月 7 日十一个国家分别与清政府签订<辛丑条约>

39th Year 第三十九年

The 60 Year Cycle 六十甲子
8 Feb 1902 - 28 Jan 1903

壬寅 Rén Yín | 阳 Yáng | 5 水 Water | 3 禄存星 Lù Cún

- 77th Cycle 4600 / 1902
- 78th Cycle 4660 / 1962
- 79th Cycle 4720 / 2022

Heavenly Stem 9 壬 Rén
Earthly Branch 3 寅虎 Tiger

Dowager Empress Cíxǐ and Emperor Guāngxù return to Běijīng from Xī'ān in 1902.
1902 年慈禧和光绪从西安返回北京.

40th Year 第四十年

The 60 Year Cycle 六十甲子
29 Jan 1903 - 15 Feb 1904

癸卯 Guǐ Mǎo | 阴 Yīn | 5 水 Water | 4 文曲星 Wén Qū

- 77th Cycle 4601 / 1903
- 78th Cycle 4661 / 1963
- 79th Cycle 4721 / 2023

Heavenly Stem 10 癸 Guǐ
Earthly Branch 4 卯兔 Rabbit

British invade Tibet December 1903; thousands of Tibetans killed. 1903 年 12 月英军入侵西藏,上千藏民被杀.

Astronomical Pencil Drawings - John Oxenham Goodman

41st Year 第四十一年
The 60 Year Cycle 六十甲子
16 Feb 1904 - 3 Feb 1905

甲辰 Jiǎ Chén | 阳 Yáng | 1 Wood | 5 廉贞星 Lián Zhēn

- 77th Cycle 4602
- 78th Cycle 4662
- 79th Cycle 4722
- 77th Cycle 1904
- 78th Cycle 1964
- 79th Cycle 2024

Heavenly Stem: 1 甲 Jiǎ
Earthly Branch: 5 辰龙 Dragon

Japanese gun used in siege of Port Arthur (Lǚ Xùn Kǒu). 1 Aug 1904 - 2 Jan 1905. Russians surrender.
1904 年日军炮轰旅顺口,俄军投降。

42nd Year 第四十二年
The 60 Year Cycle 六十甲子
4 Feb 1905 - 24 Jan 1906

乙巳 Yǐ Sì | 阴 Yīn | 1 Wood | 6 武曲星 Wǔ Qū

- 77th Cycle 4603
- 78th Cycle 4663
- 79th Cycle 4723
- 77th Cycle 1905
- 78th Cycle 1965
- 79th Cycle 2025

Heavenly Stem: 2 乙 Yǐ
Earthly Branch: 6 巳蛇 Snake

Russian artillery in Battle of Mukden (now Shěnyáng) in 1905. 26,000 Russians and 41,000 Japanese killed, Russia defeated.
1905 年奉天会战,俄军用大炮抵抗日军,最终失败。

43rd Year 第四十三年
The 60 Year Cycle 六十甲子
25 Jan 1906 - 12 Feb 1907

丙午 Bǐng Wǔ | 阳 Yáng | 2 Fire | 7 破军星 Pò Jūn

- 77th Cycle 4604
- 78th Cycle 4664
- 79th Cycle 4724
- 77th Cycle 1906
- 78th Cycle 1966
- 79th Cycle 2026

Heavenly Stem: 3 丙 Bǐng
Earthly Branch: 7 午马 Horse

Běijīng – Hànkǒu Railway 1,214 km opens 1 April 1906.
1906 年 4 月 1 日京汉铁路全线通车。

44th Year 第四十四年
The 60 Year Cycle 六十甲子
13 Feb 1907 - 1 Feb 1908

丁未 Dīng Wèi | 阴 Yīn | 2 Fire | 6 武曲星 Wǔ Qū

- 77th Cycle 4605
- 78th Cycle 4665
- 79th Cycle 4725
- 77th Cycle 1907
- 78th Cycle 1967
- 79th Cycle 2027

Heavenly Stem: 4 丁 Dīng
Earthly Branch: 8 未羊 Goat

1907 Běijīng to Paris Car Race, 14,994 km, 62 days.
1907 年北京到巴黎长途驾驶比赛,全长 14,994 公里,历时 62 天。

45th Year 第四十五年
The 60 Year Cycle 六十甲子
2 Feb 1908 - 21 Jan 1909

戊申 Wù Shēn | 阳 Yáng | 3 土 Earth | 5 廉贞星 Lián Zhēn

- 77th Cycle 4606
- 78th Cycle 4666
- 79th Cycle 4726
- 77th Cycle 1908
- 78th Cycle 1968
- 79th Cycle 2028

Heavenly Stem **5** 戊 Wù
Earthly Branch **9** 申猴 Monkey

Pǔyí becomes the Xuān Tǒng Emperor on 2 December 1908 at the age of 2 years 10 months.
1908 年 12 月 2 日溥仪继位, 改号宣统.

46th Year 第四十六年
The 60 Year Cycle 六十甲子
22 Jan 1909 - 9 Feb 1910

己酉 Jǐ Yǒu | 阴 Yīn | 3 土 Earth | 4 文曲星 Wén Qǔ

- 77th Cycle 4607
- 78th Cycle 4667
- 79th Cycle 4727
- 77th Cycle 1909
- 78th Cycle 1969
- 79th Cycle 2029

Heavenly Stem **6** 己 Jǐ
Earthly Branch **10** 酉鸡 Rooster

Naval ships Hǎi Qí and Hǎi Róng visit Hong Kong, Singapore, Jakarta and Saigon.
1909 年海圻率海容舰访问东南亚.

47th Year 第四十七年
The 60 Year Cycle 六十甲子
10 Feb 1910 - 29 Jan 1911

庚戌 Gēng Xū | 阳 Yáng | 4 金 Metal | 3 禄存星 Lù Cún

- 77th Cycle 4608
- 78th Cycle 4668
- 79th Cycle 4728
- 77th Cycle 1910
- 78th Cycle 1970
- 79th Cycle 2030

Heavenly Stem **7** 庚 Gēng
Earthly Branch **11** 戌狗 Dog

First exposition in China held in Nánjīng in 1910.
1910 年南京举办中国首次博览会.

48th Year 第四十八年
The 60 Year Cycle 六十甲子
30 Jan 1911 - 17 Feb 1912

辛亥 Xīn Hài | 阴 Yīn | 4 金 Metal | 2 巨门星 Jù Mén

- 77th Cycle 4609
- 78th Cycle 4669
- 79th Cycle 4729
- 77th Cycle 1911
- 78th Cycle 1971
- 79th Cycle 2031

Heavenly Stem **8** 辛 Xīn
Earthly Branch **12** 亥猪 Pig

1911 年 10 月 10 日辛亥革命 (1911 年 10 月 10 日 — 1912 年 2 月 12 日)之武昌起义.

Xīn Hài Revolution 10 Oct 1911 – 12 Feb 1912. Emperor Pǔyí forced to abdicate.

49th Year 第四十九年
The 60 Year Cycle 六十甲子
18 Feb 1912 - 5 Feb 1913

壬子 Rén Zǐ | 阳 Yáng | 5 Water | 1 贪狼星 Tān Láng

77th Cycle 4610
78th Cycle 4670
79th Cycle 4730

77th Cycle 1912
78th Cycle 1972
79th Cycle 2032

Yuán Shì Kǎi replaces Sun Yat-sen as Provisional President 10 Mar 1912.
1912年3月10日袁世凯接替孙中山为临时大总统。

Heavenly Stem 9 壬 Rén
Earthly Branch 1 子鼠 Rat

50th Year 第五十年
The 60 Year Cycle 六十甲子
6 Feb 1913 - 25 Jan 1914

癸丑 Guǐ Chǒu | 阴 Yīn | 5 Water | 2 巨门星 Jù Mén

77th Cycle 4611
78th Cycle 4671
79th Cycle 4731

77th Cycle 1913
78th Cycle 1973
79th Cycle 2033

July 1913 Second Revolution failed to overthrow Yuán Shì-kǎi.
1913年7月试图推翻袁世凯的二次革命失败。

Heavenly Stem 10 癸 Guǐ
Earthly Branch 2 丑牛 Ox

51st Year 第五十一年
The 60 Year Cycle 六十甲子
26 Jan 1914 - 13 Feb 1915

甲寅 Jiǎ Yín | 阳 Yáng | 1 Wood | 3 禄存星 Lù Cún

77th Cycle 4612
78th Cycle 4672
79th Cycle 4732

77th Cycle 1914
78th Cycle 1974
79th Cycle 2034

Japanese siege of Qīngdǎo 31 Oct 1914 – 7 Nov 1914. Germans surrender.
1914年10月31日—1914年11月7日,日军攻打青岛,德军投降。

Heavenly Stem 1 甲 Jiǎ
Earthly Branch 3 寅虎 Tiger

52nd Year 第五十二年
The 60 Year Cycle 六十甲子
14 Feb 1915 - 2 Feb 1916

乙卯 Yǐ Mǎo | 阴 Yīn | 1 Wood | 4 文曲星 Wén Qū

77th Cycle 4613
78th Cycle 4673
79th Cycle 4733

77th Cycle 1915
78th Cycle 1975
79th Cycle 2035

National Protection War begins 25 Dec 1915. Emperor Yuán Shì-kǎi forced to abdicate on 20 March 1916.
1915年12月25日护国运动开始,次年袁世凯被迫下台。

Heavenly Stem 2 乙 Yǐ
Earthly Branch 4 卯兔 Rabbit

53rd Year
第五十三年

The 60 Year Cycle
六十甲子
3 Feb 1916 - 22 Jan 1917

丙辰 Bīng Chén | 阳 Yáng | **2** Fire | 5 廉贞星 ⭐ Lián Zhēn

- 77th Cycle 4614
- 78th Cycle 4674
- 79th Cycle 4734
- 77th Cycle 1916
- 78th Cycle 1976
- 79th Cycle 2036

Troop transport Xīn Yù sinks after collision with Hǎi Róng in fog. 1,000 soldiers drown.
1916 年 4 月 20 日海容舰撞沉新裕轮，船上 1000 多名官兵溺死。

Heavenly Stem **3** 丙 Bǐng
Earthly Branch **5** 辰龙 Dragon

54th Year
第五十四年

The 60 Year Cycle
六十甲子
23 Jan 1917 - 10 Feb 1918

丁巳 Dīng Sì | 阴 Yīn | **2** Fire | 6 武曲星 ⭐ Wǔ Qū

- 77th Cycle 4615
- 78th Cycle 4675
- 79th Cycle 4735
- 77th Cycle 1917
- 78th Cycle 1977
- 79th Cycle 2037

Xuān Tǒng Emperor Pǔyí restored to throne 1 July 1917 but dethroned on 12 July.
1917 年 7 月 1 日张勋策划溥仪复辟，7 月 12 日以失败告终。

Heavenly Stem **4** 丁 Dīng
Earthly Branch **6** 巳蛇 Snake

55th Year
第五十五年

The 60 Year Cycle
六十甲子
11 Feb 1918 - 31 Jan 1919

戊午 Wù Wǔ | 阳 Yáng | **3** Earth | 7 破军星 ⭐ Pò Jūn

- 77th Cycle 4616
- 78th Cycle 4676
- 79th Cycle 4736
- 77th Cycle 1918
- 78th Cycle 1978
- 79th Cycle 2038

中国军队在海参崴
Chinese troops in Vladivostok (Hǎi Shēn Wǎi)
April 1918, 3000 Chinese troops join Allied Intervention in Siberia.
1918 年 4 月中国 3000 名士兵参加联合干涉军，出兵西伯利亚。

Heavenly Stem **5** 戊 Wù
Earthly Branch **7** 午马 Horse

56th Year
第五十六年

The 60 Year Cycle
六十甲子
1 Feb 1919 - 19 Feb 1920

己未 Jǐ Wèi | 阴 Yīn | **3** Earth | 6 武曲星 ⭐ Wǔ Qū

- 77th Cycle 4617
- 78th Cycle 4677
- 79th Cycle 4737
- 77th Cycle 1919
- 78th Cycle 1979
- 79th Cycle 2039

May 4th Movement 1919. Anti-Japanese demonstrations.
1919 年反日爱国的五四运动游行。

Heavenly Stem **6** 己 Jǐ
Earthly Branch **8** 未羊 Goat

Wújí (Infinity), Tàijí (the Supreme Ultimate), 2 Forms and 4 Phenomena 无极，太极，两仪和四象

Wújí 无极 (無極) is the limitless cosmic first principle which is non-polar and which predates the existential and material world. Tàijí is said to be the "Great Primal Beginning". Wújí is represented by a circle while Tàijí is portrayed as a circle divided into dark and light sections known as Yīn 阴 and Yáng 阳. Tàijí is thus the Yīn-Yáng principle of bipolarity from which existence evolved. Yáng is represented by an unbroken line while Yīn is symbolized by a broken line and these two are known as the Two Forms (Liǎng Yí 两仪). They in turn produce the Four Phenomena (Sì Xiàng 四象) : Great Yáng (太阳 Tài Yáng), Lesser Yīn (少阴 Shǎo Yīn), Great Yīn (太阴 Tài Yīn), Lesser Yáng (少阳 Shǎo Yáng) which are related to the 4 directions and 4 seasons.

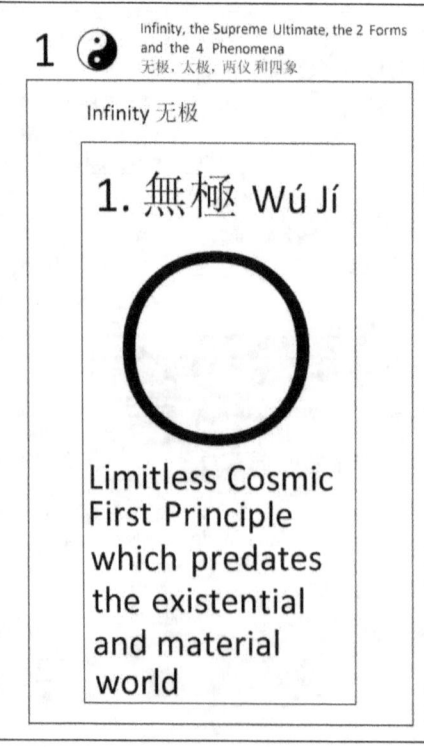

3 ☯ Infinity, the Supreme Ultimate, the 2 Forms and the 4 Phenomena 无极，太极，两仪 和四象 Chemical Elements **107** The Supreme Ultimate 太极  The Great Absolute, the Great Primal Beginning, the Yin-Yang Principle of Bipolarity	**4** ☯ Infinity, the Supreme Ultimate, the 2 Forms and the 4 Phenomena 无极，太极，两仪 和四象 Chemical Elements **2, 42, 60, 89, 91** 南 **S** Shang Suns **3** Sun, Planets & Moons **0** Asteroids & Moons, Comets **1, 10, 93** Tarot 囧 **19** **2** 火 Fire  Yáng：Sun, Fire, Male, South, Summer
5 ☯ Infinity, the Supreme Ultimate, the 2 Forms and the 4 Phenomena 无极，太极，两仪 和四象 Chemical Elements **34, 59** 北 **N** Planets & Moons **3/1** Tarot 囧 **2, 18** **5** 水 Water  Yīn：Moon, Water, Female, North, Winter	**6** ☯ Infinity, the Supreme Ultimate, the 2 Forms and the 4 Phenomena 无极，太极，两仪 和四象 南 **S** **2** 火 Fire  Three Pagodas at Dàlǐ in Yúnnán Province 云南大理三塔寺 Tài Yáng：Sun, South, Summer, Midday

7 ☯ Infinity, the Supreme Ultimate, the 2 Forms and the 4 Phenomena
无极，太极，两仪 和四象

4 金
Metal

西 W 西 秋 **7.** 少陰 ⚎

Dragon Subduing Pagoda, Red Hill, Ürümqi
乌鲁木齐红山镇龙塔

Shǎo Yīn: Setting Sun, West, Autumn

---

8 ☯ Infinity, the Supreme Ultimate, the 2 Forms and the 4 Phenomena
无极，太极，两仪 和四象

5 水
Water

北 N 北 冬 **8.** 太陰 ⚏

Harbin, Songhua River
哈尔滨松花江

Tài Yīn: Moon, Night, North, Winter, Ice, Snow

---

9 ☯ Infinity, the Supreme Ultimate, the 2 Forms and the 4 Phenomena
无极，太极，两仪 和四象

1 木
Wood

东 E 东 春 **9.** 少陽 ⚍

上海东方明珠塔

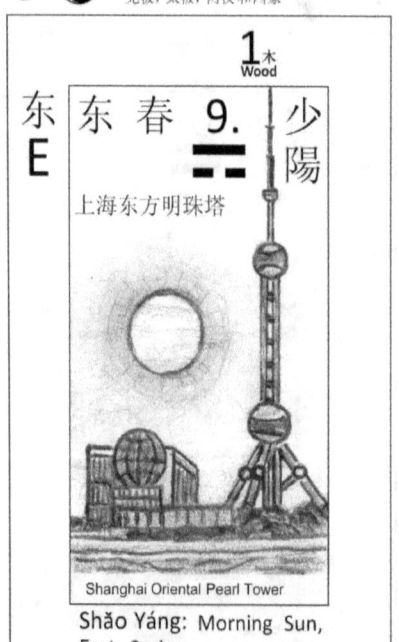

Shanghai Oriental Pearl Tower

Shǎo Yáng: Morning Sun, East, Spring

---

10 ☯ Infinity, the Supreme Ultimate, the 2 Forms and the 4 Phenomena
无极，太极，两仪 和四象

10. The Eight Trigrams and the Yīn Yáng Symbol 八卦和阴阳鱼

Chemical Elements 107

Marsh 泽 Duì — Heaven 天 Qián — Wind 风 Xùn
Fire 火 Lí — Kǎn Water 坎水
Zhèn 震 Thunder 雷 — Kūn 坤 地 Earth — Gèn 艮 山 Mountain

Earlier Heaven Bā Guà 先天八卦

There are two versions of The Eight Diagrams or Trigrams (八卦 Bā Guà)

The Primordial or Earlier Heaven Bā Guà (先天八卦 Xiān Tiān Bā Guà) believed to have been invented by Fú Xī 伏羲 (2953 BC-2852 BC).

The Later Heaven Bā Guà (后天八卦 Hòu Tiān Bā Guà) supposedly invented by King Wen (周文王 Zhōu Wén Wáng, 1152 BC-1056 BC) the founder of the Zhōu Dynasty

Both versions have trigrams entitled Heaven, Wind, Water, Mountain, Earth, Thunder, Fire and Marsh which are related to the 5 Elements, the 8 directions and the 4 seasons. However, their relationship with the Seasons and Directions is different in the Later Heaven sequence.

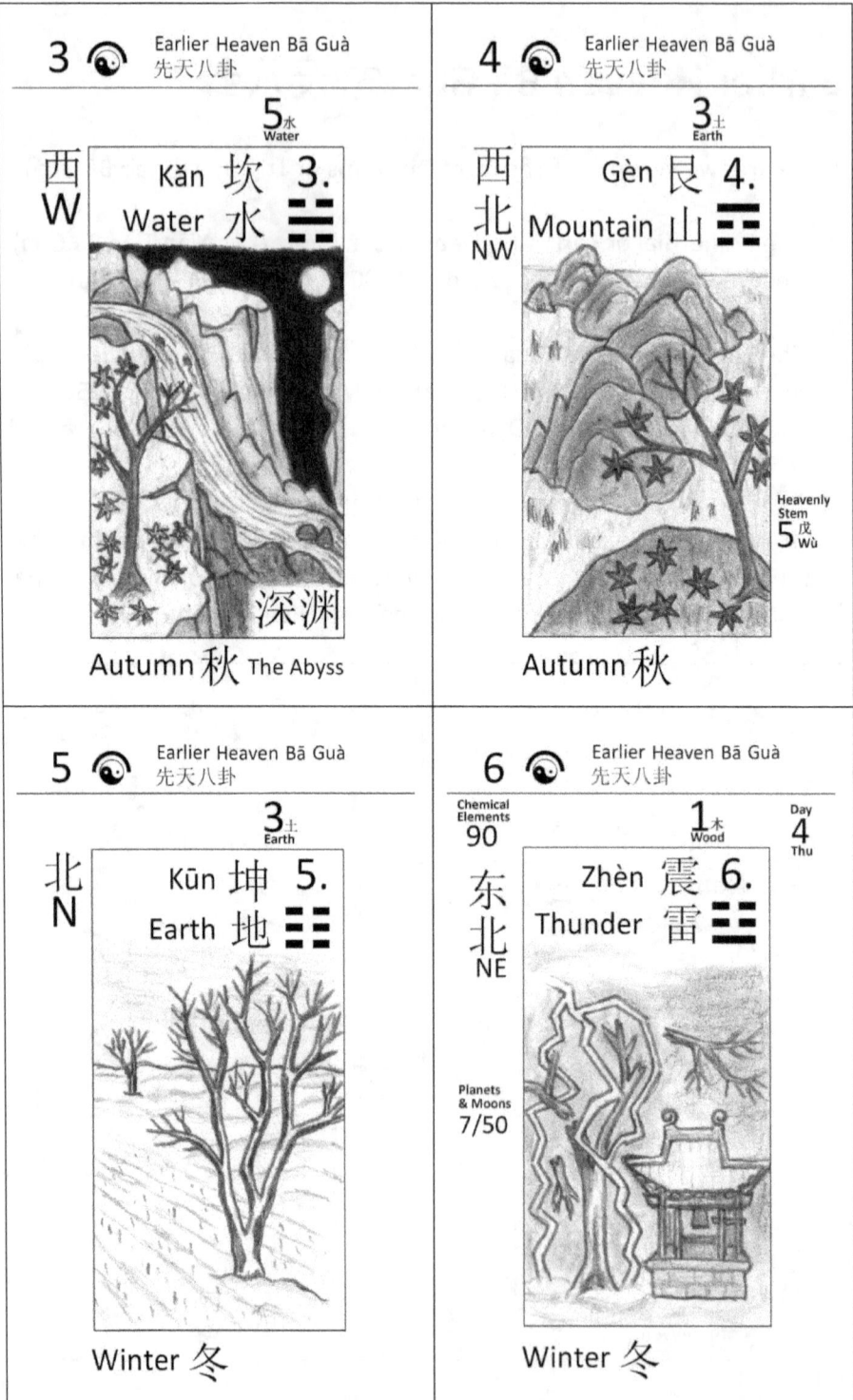

Later Heaven Ba Gua 后天八卦

There are two versions of the Eight Diagrams or Trigrams (八卦 Bā Guà).

The Primordial or Earlier Heaven Bā Guà (先天八卦 Xiān Tiān Bā Guà) believed to have been invented by Fú Xī 伏羲 (2953 BC-2852 BC).

The Later Heaven Bā Guà (后天八卦 Hòu Tiān Bā Guà) supposedly invented by King Wen (周文王 Zhōu Wén Wáng, 1152 BC-1056 BC) the founder of the Zhōu Dynasty.

Both versions have trigrams entitled Heaven, Wind, Water, Mountain, Earth, Thunder, Fire and Marsh which are related to the 5 Elements, the 8 directions and the 4 seasons. However, Later Heaven has a different relationship with the Seasons and Directions and has an animal for each of the 8 trigrams some of which are related to the 28 Lunar Mansions.

Martial Arts Bā Guà 尹氏八卦掌 founded by Yǐn Fú 尹福

The Yǐn Shì Bā Guà Zhǎng 尹氏八卦掌 was founded by Yǐn Fú 尹福 and passed down by Xiè Pèi Qǐ 解佩启. Martial Arts Bā Guà are based on observations of the movements of animals. They can be arranged in different sequences depending on which order the movements are to follow. Some of the animals are related to animals in the 28 Lunar Mansions, the 12 Earthly Branches, the 88 Constellations, the Planets and Moons, the Asteroids and moons and the Tarot cards.

The 64 Hexagrams of the Classic of Changes 易经的六十四卦

The Yì Jīng 易经, also known as the Zhōu Yì 周易, is called the *Classic of Changes* or *Book of Changes* in English. It contains ancient cosmic principles and has had a strong influence on Chinese philosophy. However, it is also used at the popular level as a tool for divination and fortune telling. A hexagram is composed of two trigrams, one placed on top of the other. The broken lines represents Yīn 阴 the receptive principle while the unbroken lines represent the creative principle Yáng 阳. The forces of Yīn and Yáng are complementary and when one increases, the other decreases causing change. A hexagram is therefore a stack of six broken or unbroken lines and there are 64 possibilities for arranging its lines. The three lower lines (lower trigram) of a hexagram are related to internal change whereas the three upper lines (upper trigram) represent external change.

Astronomical Pencil Drawings - John Oxenham Goodman

---

1  The 64 Hexagrams 六十四卦

4 金
Metal

1 乾 Qián
FORCE
The Creative
Heaven,
creative, strong action

☰ 乾 Qián = 天 Heaven
FORCE, The Creative, strong, expansive energy, the sky
Element: 金 **Metal 4**

☰ 乾 Qián = 天 Heaven
FORCE, The Creative, strong, expansive energy, the sky
Element: 金 **Metal 4**

---

2  The 64 Hexagrams 六十四卦

3 土
Earth

2 坤 Kūn
FIELD
The Receptive
Earth,
acquiescence

☷ 坤 Kūn = 地 Earth,
FIELD, The Receptive, that which yields, receptive energy
Element: 土 **Earth 3**

☷ 坤 Kūn = 地 Earth,
FIELD, The Receptive, that which yields, receptive energy
Element: 土 **Earth 3**

---

3  The 64 Hexagrams 六十四卦

3 屯 Zhūn
SPROUTING
Difficulty at the Beginning,
gathering support, hoarding

☵ 坎 Kǎn = 水 Water
GORGE, The Abysmal, dangerous, rapid rivers, the abyss, the moon
Element: 水 **Water 5**

☳ 震 Zhèn = 雷 Thunder
SHAKE, The Arousing, inciting, movement, initiative, revolution
Element: 木 **Wood 1**

---

4  The 64 Hexagrams 六十四卦

4 蒙 Méng
ENVELOPING
Youthful Folly, the young shoot, discovering

☶ 艮 Gèn = 山 Mountain
BOUND, stillness, keeping still, resting, unable to be moved
Element: 土 **Earth 3**

☵ 坎 Kǎn = 水 Water
GORGE, The Abysmal, dangerous, rapid rivers, the abyss, the moon
Element: 水 **Water 5**

5 ▤ The 64 Hexagrams 六十四卦

5 需 Xū
ATTENDING
Waiting, moistened, arriving

☵ 坎 Kǎn = 水 Water
GORGE, The Abysmal, dangerous, rapid rivers, the abyss, the moon
Element: 水 Water 5

☰ 乾 Qián = 天 Heaven
FORCE, The Creative, strong, expansive energy, the sky
Element: 金 Metal 4

6 ▤ The 64 Hexagrams 六十四卦

6 訟 Sòng
ARGUING
Conflict, lawsuit

☰ 乾 Qián = 天 Heaven
FORCE, The Creative, strong, expansive energy, the sky
Element: 金 Metal 4

☵ 坎 Kǎn = 水 Water
GORGE, The Abysmal, dangerous, rapid rivers, the abyss, the moon
Element: 水 Water 5

7 ▤ The 64 Hexagrams 六十四卦

7 師 Shī
LEADING
The Army

☷ 坤 Kūn = 地 Earth,
FIELD, The Receptive, that which yields, receptive energy
Element: 土 Earth 3

☵ 坎 Kǎn = 水 Water
GORGE, The Abysmal, dangerous, rapid rivers, the abyss, the moon
Element: 水 Water 5

8 ▤ The 64 Hexagrams 六十四卦

8 比 Bǐ
GROUPING
Holding Together, alliance

☵ 坎 Kǎn = 水 Water
GORGE, The Abysmal, dangerous, rapid rivers, the abyss, the moon
Element: 水 Water 5

☷ 坤 Kūn = 地 Earth,
FIELD, The Receptive, that which yields, receptive energy
Element: 土 Earth 3

9 ▦ The 64 Hexagrams 六十四卦

9 小畜 Xiǎo Chù
SMALL ACCUMULATING

Small Taming, the taming power of the small, small harvest

☴ 巽 Xùn = 風 Wind
GROUND, the gentle, gentle penetration, entrance, flexibility
   Element: 木 Wood 1

☰ 乾 Qián = 天 Heaven
FORCE, The Creative, strong, expansive energy, the sky
   Element: 金 Metal 4

---

10 ▦ The 64 Hexagrams 六十四卦

4 金
Metal

10 履 Lǚ
TREADING

Treading (conduct), continuing

☰ 乾 Qián = 天 Heaven
FORCE, The Creative, strong, expansive energy, the sky
   Element: 金 Metal 4

☱ 兌 Duì = 澤 Marsh, Swamp
OPEN, the joyous, pleasure, satisfaction, stagnation
   Element: 金 Metal 4

---

11 ▦ The 64 Hexagrams 六十四卦

11 泰 Tài
PERVADING

Peace, greatness

☷ 坤 Kūn = 地 Earth,
FIELD, The Receptive, that which yields, receptive energy
   Element: 土 Earth 3

☰ 乾 Qián = 天 Heaven
FORCE, The Creative, strong, expansive energy, the sky
   Element: 金 Metal 4

---

12 ▦ The 64 Hexagrams 六十四卦

12 否 Pǐ
OBSTRUCTION

Standstill, stagnation, selfish persons

☰ 乾 Qián = 天 Heaven
FORCE, The Creative, strong, expansive energy, the sky
   Element: 金 Metal 4

☷ 坤 Kūn = 地 Earth,
FIELD, The Receptive, that which yields, receptive energy
   Element: 土 Earth 3

13 The 64 Hexagrams 六十四卦

13 同人
Tóng Rén
CONCORDING PEOPLE

Fellowship, gathering people, partnership

☰ 乾 **Qián** = 天 **Heaven**
FORCE, The Creative, strong, expansive energy, the sky
Element: 金 **Metal 4**

☲ 離 **Lí** = 火 **Fire**
RADIANCE, The Clinging, rapid movement, clarity, adaptable, the sun
Element: 火 **Fire 2**

14 The 64 Hexagrams 六十四卦

14 大有
Dà Yǒu
GREAT POSSESSING

Great Possession, possession in great measure

☲ 離 **Lí** = 火 **Fire**
RADIANCE, The Clinging, rapid movement, clarity, adaptable, the sun
Element: 火 **Fire 2**

☰ 乾 **Qián** = 天 **Heaven**
FORCE, The Creative, strong, expansive energy, the sky
Element: 金 **Metal 4**

15 The 64 Hexagrams 六十四卦

3 土 Earth

15 謙 **Qiān**
HUMBLING
Modesty

☷ 坤 **Kūn** = 地 **Earth**,
FIELD, The Receptive, that which yields, receptive energy
Element: 土 **Earth 3**

☶ 艮 **Gèn** = 山 **Mountain**
BOUND, stillness, keeping still, resting, unable to be moved
Element: 土 **Earth 3**

16 The 64 Hexagrams 六十四卦

16 豫 **Yù**
PROVIDING-FOR
Enthusiasm, excess

☳ 震 **Zhèn** = 雷 **Thunder**
SHAKE, The Arousing, inciting, movement, initiative, revolution
Element: 木 **Wood 1**

☷ 坤 **Kūn** = 地 **Earth**,
FIELD, The Receptive, that which yields, receptive energy
Element: 土 **Earth 3**

17 ▦ The 64 Hexagrams 六十四卦

17 隨 Suí
FOLLOWING
Following

☱ 兌 Duì = 澤 **Marsh, Swamp**
OPEN, the joyous, pleasure, satisfaction, stagnation
　　Element: 金 **Metal 4**

☳ 震 Zhèn = 雷 **Thunder**
SHAKE, The Arousing, inciting, movement, initiative, revolution
　　Element: 木 **Wood 1**

18 ▦ The 64 Hexagrams 六十四卦

18 蠱 Gǔ
CORRECTING
Work on the Decayed, work on what has been spoiled. *Gu* is a poison used in Chinese witchcraft.

☶ 艮 Gèn = 山 **Mountain**
BOUND, stillness, keeping still, resting, unable to be moved
　　Element: 土 **Earth 3**

☴ 巽 Xùn = 風 **Wind**
GROUND, the gentle, gentle penetration, entrance, flexibility
　　Element: 木 **Wood 1**

19 ▦ The 64 Hexagrams 六十四卦

19 臨 Lín
NEARING
Approach

☷ 坤 Kūn = 地 **Earth**
FIELD, The Receptive, that which yields, receptive energy
　　Element: 土 **Earth 3**

☱ 兌 Duì = 澤 **Marsh, Swamp**
OPEN, the joyous, pleasure, satisfaction, stagnation
　　Element: 金 **Metal 4**

20 ▦ The 64 Hexagrams 六十四卦

20 觀 Guān
VIEWING
Contemplation (view)

☴ 巽 Xùn = 風 **Wind**
GROUND, the gentle, gentle penetration, entrance, flexibility
　　Element: 木 **Wood 1**

☷ 坤 Kūn = 地 **Earth**
FIELD, The Receptive, that which yields, receptive energy
　　Element: 土 **Earth 3**

21 ䷔ The 64 Hexagrams 六十四卦

21 噬嗑
Shì Kè
GNAWING Biting
Through, bite, biting and chewing

☲ 離 Lí = 火 Fire
RADIANCE, The Clinging, rapid movement, clarity, adaptable, the sun
　　Element: 火 Fire 2

☳ 震 Zhèn = 雷 Thunder
SHAKE, The Arousing, inciting, movement, initiative, revolution
　　Element: 木 Wood 1

22 ䷕ The 64 Hexagrams 六十四卦

22 賁 Bì
ADORNING
Grace, luxuriance, embellishing

☶ 艮 Gèn = 山 Mountain
BOUND, stillness, keeping still, resting, unable to be moved
　　Element: 土 Earth 3

☲ 離 Lí = 火 Fire
RADIANCE, The Clinging, rapid movement, clarity, adaptable, the sun
　　Element: 火 Fire 2

23 ䷖ The 64 Hexagrams 六十四卦

3 土 Earth

23 剝 Bō
STRIPPING
Splitting Apart, flaying

☶ 艮 Gèn = 山 Mountain
BOUND, stillness, keeping still, resting, unable to be moved
　　Element: 土 Earth 3

☷ 坤 Kūn = 地 Earth,
FIELD, The Receptive, that which yields, receptive energy
　　Element: 土 Earth 3

24 ䷗ The 64 Hexagrams 六十四卦

24 復 Fù
RETURNING
Return, the turning point

☷ 坤 Kūn = 地 Earth,
FIELD, The Receptive, that which yields, receptive energy
　　Element: 土 Earth 3

☳ 震 Zhèn = 雷 Thunder
SHAKE, The Arousing, inciting, movement, initiative, revolution
　　Element: 木 Wood 1

25 ䷘ The 64 Hexagrams 六十四卦

25 無妄 Wú Wàng
WITHOUT EMBROILING
Innocence (the unexpected)

☰ 乾 Qián = 天 **Heaven**
FORCE, The Creative, strong, expansive energy, the sky
Element: 金 **Metal 4**

☳ 震 Zhèn = 雷 **Thunder**
SHAKE, The Arousing, inciting, movement, initiative, revolution
Element: 木 **Wood 1**

26 ䷙ The 64 Hexagrams 六十四卦

Dà Chù
GREAT ACCUMULATING
Great Taming, the taming power of the great, great storage, potential energy

☶ 艮 Gèn = 山 **Mountain**
BOUND, stillness, keeping still, resting, unable to be moved
Element: 土 **Earth 3**

☰ 乾 Qián = 天 **Heaven**
FORCE, The Creative, strong, expansive energy, the sky
Element: 金 **Metal 4**

27 ䷚ The 64 Hexagrams 六十四卦

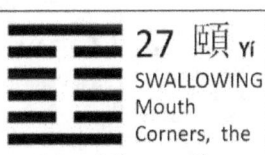
SWALLOWING
Mouth Corners, the corners of the mouth (providing nourishment), jaws, comfort/security

☶ 艮 Gèn = 山 **Mountain**
BOUND, stillness, keeping still, resting, unable to be moved
Element: 土 **Earth 3**

☳ 震 Zhèn = 雷 **Thunder**
SHAKE, The Arousing, inciting, movement, initiative, revolution
Element: 木 **Wood 1**

28 ䷛ The 64 Hexagrams 六十四卦

Dà Guò
GREAT EXCEEDING
Great Preponderance, preponderance of the great, great surpassing, critical mass

☱ 兌 Duì = 澤 **Marsh, Swamp**
OPEN, the joyous, pleasure, satisfaction, stagnation
Element: 金 **Metal 4**

☴ 巽 Xùn = 風 **Wind**
GROUND, the gentle, gentle penetration, entrance, flexibility
Element: 木 **Wood 1**

29 The 64 Hexagrams 六十四卦

5 水 Water

29 坎 Kǎn
GORGE
Abysmal
Water,
repeated entrapment,
darkness

☵ 坎 Kǎn = 水 Water
GORGE, The Abysmal,
dangerous rapid rivers,
the abyss, the moon
Element: 水 Water 5

☵ 坎 Kǎn = 水 Water
GORGE, The Abysmal,
dangerous rapid rivers,
the abyss, the moon
Element: 水 Water 5

30 The 64 Hexagrams 六十四卦

2 火 Fire

30 離 Lí
RADIANCE
The Clinging
Fire, the net

☲ 離 Lí = 火 Fire
RADIANCE, The Clinging,
rapid movement, clarity,
adaptable, the sun
Element: 火 Fire 2

☲ 離 Lí = 火 Fire
RADIANCE, The Clinging,
rapid movement, clarity,
adaptable, the sun
Element: 火 Fire 2

31 The 64 Hexagrams 六十四卦

31 咸 Xián
CONJOINING
Influence,
Influence
(wooing), feelings

☱ 兌 Duì = 澤 Marsh, Swamp
OPEN, the joyous, pleasure,
satisfaction, stagnation
Element: 金 Metal 4

☶ 艮 Gèn = 山 Mountain
BOUND, stillness, keeping
still, resting, unable to be
moved
Element: 土 Earth 3

32 The 64 Hexagrams 六十四卦

1 木 Wood

32 恆 Héng
PERSEVERING
Duration,
constancy,
perseverance

☳ 震 Zhèn = 雷 Thunder
SHAKE, The Arousing,
inciting, movement,
initiative, revolution
Element: 木 Wood 1

☴ 巽 Xùn = 風 Wind
GROUND, the gentle, gentle
penetration, entrance,
flexibility
Element: 木 Wood 1

33 ䷠ The 64 Hexagrams 六十四卦

 33 遯 Dùn
RETIRING
Retreat, withdrawing, yielding

☰ 乾 Qián = 天 Heaven
FORCE, The Creative, strong, expansive energy, the sky
Element: 金 Metal 4

☶ 艮 Gèn = 山 Mountain
BOUND, stillness, keeping still, resting, unable to be moved
Element: 土 Earth 3

34 ䷡ The 64 Hexagrams 六十四卦

 34 大壯
Dà Zhuàng
GREAT INVIGORATING
Great Power, the power of the great, great maturity

☳ 震 Zhèn = 雷 Thunder
SHAKE, The Arousing, inciting, movement, initiative, revolution
Element: 木 Wood 1

☰ 乾 Qián = 天 Heaven
FORCE, The Creative, strong, expansive energy, the sky
Element: 金 Metal 4

35 ䷢ The 64 Hexagrams 六十四卦

 35 晉 Jìn
PROSPERING
Progress

☲ 離 Lí = 火 Fire
RADIANCE, The Clinging, rapid movement, clarity, adaptable, the sun
Element: 火 Fire 2

☷ 坤 Kūn = 地 Earth,
FIELD, The Receptive, that which yields, receptive energy
Element: 土 Earth 3

36 ䷣ The 64 Hexagrams 六十四卦

 36 明夷
Míng Yí
BRIGHTNESS HIDING
Darkening of the Light, brilliance injured, intelligence hidden

☷ 坤 Kūn = 地 Earth,
FIELD, The Receptive, that which yields, receptive energy
Element: 土 Earth 3

☲ 離 Lí = 火 Fire
RADIANCE, The Clinging, rapid movement, clarity, adaptable, the sun
Element: 火 Fire 2

37 ䷤ The 64 Hexagrams 六十四卦

37 家人 Jiā Rén
DWELLING PEOPLE
The Family, the clan, family members

☴ **Xùn** = 風 **Wind**
GROUND, the gentle, gentle penetration, entrance, flexibility
   Element: 木 Wood 1

☲ **Lí** = 火 **Fire**
RADIANCE, The Clinging, rapid movement, clarity, adaptable, the sun
   Element: 火 Fire 2

38 ䷥ The 64 Hexagrams 六十四卦

38 睽 Kuí
POLARISING
Opposition, division, divergence, perversion

☲ **Lí** = 火 **Fire**
RADIANCE, The Clinging, rapid movement, clarity, adaptable, the sun
   Element: 火 Fire 2

☱ **Duì** = 澤 **Marsh, Swamp**
OPEN, the joyous, pleasure, satisfaction, stagnation
   Element: 金 Metal 4

39 ䷦ The 64 Hexagrams 六十四卦

39 蹇 Jiǎn
LIMPING
Obstruction

☵ **Kǎn** = 水 **Water**
GORGE, The Abysmal, dangerous, rapid rivers, the abyss, the moon
   Element: 水 Water 5

☶ **Gèn** = 山 **Mountain**
BOUND, stillness, keeping still, resting, unable to be moved
   Element: 土 Earth 3

40 ䷧ The 64 Hexagrams 六十四卦

40 解 Xiè
TAKING-APART
Deliverance, untangled, solution

☳ **Zhèn** = 雷 **Thunder**
SHAKE, The Arousing, inciting, movement, initiative, revolution
   Element: 木 Wood 1

☵ **Kǎn** = 水 **Water**
GORGE, The Abysmal, dangerous, rapid rivers, the abyss, the moon
   Element: 水 Water 5

41 ䷨ The 64 Hexagrams 六十四卦

41 損 Sǔn
DIMINISHING
Decrease

☶ 艮 Gèn = 山 Mountain
BOUND, stillness, keeping still, resting, unable to be moved
Element: 土 **Earth 3**

☱ 兌 Duì = 澤 Marsh, Swamp
OPEN, the joyous, pleasure, satisfaction, stagnation
Element: 金 **Metal 4**

42 ䷩ The 64 Hexagrams 六十四卦

1 木 Wood

42 益 Yì
AUGMENTING
Increase

☴ 巽 Xùn = 風 Wind
GROUND, the gentle, gentle penetration, entrance, flexibility
Element: 木 **Wood 1**

☳ 震 Zhèn = 雷 Thunder
SHAKE, The Arousing, inciting, movement, initiative, revolution
Element: 木 **Wood 1**

43 ䷪ The 64 Hexagrams 六十四卦

4 金 Metal

43 夬 Guài
PARTING
Breakthrough, separation, displacement

☱ 兌 Duì = 澤 Marsh, Swamp
OPEN, the joyous, pleasure, satisfaction, stagnation
Element: 金 **Metal 4**

☰ 乾 Qián = 天 Heaven
FORCE, The Creative, strong, expansive energy, the sky
Element: 金 **Metal 4**

44 ䷫ The 64 Hexagrams 六十四卦

44 姤 Gòu
COUPLING
Coming to Meet, meeting

☰ 乾 Qián = 天 Heaven
FORCE, The Creative, strong, expansive energy, the sky
Element: 金 **Metal 4**

☴ 巽 Xùn = 風 Wind
GROUND, the gentle, gentle penetration, entrance, flexibility
Element: 木 **Wood 1**

45 ䷬ The 64 Hexagrams 六十四卦

45 萃 Cuì
CLUSTERING Gathering Together, association, companionship

☱ 兑 Duì = 澤 Marsh, Swamp
OPEN, the joyous, pleasure, satisfaction, stagnation
Element: 金 Metal 4

☷ 坤 Kūn = 地 Earth,
FIELD, The Receptive, that which yields, receptive energy
Element: 土 Earth 3

46 ䷭ The 64 Hexagrams 六十四卦

46 升 Shēng
ASCENDING Pushing Upward, growing upward

☷ 坤 Kūn = 地 Earth,
FIELD, The Receptive, that which yields, receptive energy
Element: 土 Earth 3

☴ 巽 Xùn = 風 Wind
GROUND, the gentle, gentle penetration, entrance, flexibility
Element: 木 Wood 1

47 ䷮ The 64 Hexagrams 六十四卦

47 困 Kùn
CONFINING Oppression, exhaustion, entangled

☱ 兑 Duì = 澤 Marsh, Swamp
OPEN, the joyous, pleasure, satisfaction, stagnation
Element: 金 Metal 4

☵ 坎 Kǎn = 水 Water
GORGE, The Abysmal, dangerous, rapid rivers, the abyss, the moon
Element: 水 Water 5

48 ䷯ The 64 Hexagrams 六十四卦

48 井 Jǐng
WELLING The Well, replenishing, renewal

☵ 坎 Kǎn = 水 Water
GORGE, The Abysmal, dangerous, rapid rivers, the abyss, the moon
Element: 水 Water 5

☴ 巽 Xùn = 風 Wind
GROUND, the gentle, gentle penetration, entrance, flexibility
Element: 木 Wood 1

49 ䷰ The 64 Hexagrams 六十四卦

49 革 Gé
SKINNING
Revolution, moulting, abolishing the old

☱ **Duì** = 澤 **Marsh, Swamp**
OPEN, the joyous, pleasure, satisfaction, stagnation
   Element: 金 **Metal 4**

☲ **Lí** = 火 **Fire**
RADIANCE, The Clinging, rapid movement, clarity, adaptable, the sun
   Element: 火 **Fire 2**

50 ䷱ The 64 Hexagrams 六十四卦

50 鼎 Dǐng
HOLDING
The Cauldron, establishing the new

☲ **Lí** = 火 **Fire**
RADIANCE, The Clinging, rapid movement, clarity, adaptable, the sun
   Element: 火 **Fire 2**

☴ **Xùn** = 風 **Wind**
GROUND, the gentle, gentle penetration, entrance, flexibility
   Element: 木 **Wood 1**

51 ䷲ The 64 Hexagrams 六十四卦

1 木 Wood

51 震
Zhèn
SHAKE
The Arousing
Thunder, the arousing, shock, thunder

☳ **Zhèn** - 雷 **Thunder**
SHAKE, The Arousing, inciting, movement, initiative, revolution
   Element: 木 **Wood 1**

☳ **Zhèn** = 雷 **Thunder**
SHAKE, The Arousing, inciting, movement, initiative, revolution
   Element: 木 **Wood 1**

52 ䷳ The 64 Hexagrams 六十四卦

3 土 Earth

52 艮 Gèn
BOUND
The Keeping Still, mountain, immobility

☶ **Gèn** = 山 **Mountain**
BOUND, stillness, keeping still, resting, unable to be moved
   Element: 土 **Earth 3**

☶ **Gèn** = 山 **Mountain**
BOUND, stillness, keeping still, resting, unable to be moved
   Element: 土 **Earth 3**

53 ䷴ The 64 Hexagrams 六十四卦

53 漸 Jiàn
INFILTRATING
Development, gradual progress, advancement

巽 Xùn = 風 **Wind**
GROUND, the gentle, gentle penetration, entrance, flexibility
   Element: 木 **Wood 1**

艮 Gèn = 山 **Mountain**
BOUND, stillness, keeping still, resting, unable to be moved
   Element: 土 **Earth 3**

54 ䷵ The 64 Hexagrams 六十四卦

54 歸妹 Guī Mèi
CONVERTING THE MAIDEN
The Marrying Maiden, marrying, returning maiden

震 Zhèn = 雷 **Thunder**
SHAKE, The Arousing, inciting, movement, initiative, revolution
   Element: 木 **Wood 1**

兌 Duì = 澤 **Marsh, Swamp**
OPEN, the joyous, pleasure, satisfaction, stagnation
   Element: 金 **Metal 4**

55 ䷶ The 64 Hexagrams 六十四卦

55 豐 Fēng
ABOUNDING
Abundance, fullness

震 Zhèn = 雷 **Thunder**
SHAKE, The Arousing, inciting, movement, initiative, revolution
   Element: 木 **Wood 1**

離 Lí = 火 **Fire**
RADIANCE, The Clinging, rapid movement, clarity, adaptable, the sun
   Element: 火 **Fire 2**

56 ䷷ The 64 Hexagrams 六十四卦

56 旅 Lǚ
SOJOURNING
The Wanderer traveling

離 Lí = 火 **Fire**
RADIANCE, The Clinging, rapid movement, clarity, adaptable, the sun
   Element: 火 **Fire 2**

艮 Gèn = 山 **Mountain**
BOUND, stillness, keeping still, resting, unable to be moved
   Element: 土 **Earth 3**

57 ▦ The 64 Hexagrams 六十四卦

1 木 Wood

 **57 巽 Xùn**
GROUND
Gentle Wind, the gentle penetrating wind, the gentle

☴ 巽 Xùn = 風 Wind
GROUND, the gentle, gentle penetration, entrance, flexibility
　　Element: 木 Wood 1

☴ 巽 Xùn = 風 Wind
GROUND, the gentle, gentle penetration, entrance, flexibility
　　Element: 木 Wood 1

58 ▦ The 64 Hexagrams 六十四卦

4 金 Metal

 **58 兌 Duì**
OPEN
The Joyous lake, the joyous

☱ 兌 Duì = 澤 Marsh, Swamp
OPEN, the joyous, pleasure, satisfaction, stagnation
　　Element: 金 Metal 4

☱ 兌 Duì = 澤 Marsh, Swamp
OPEN, the joyous, pleasure, satisfaction, stagnation
　　Element: 金 Metal 4

59 ▦ The 64 Hexagrams 六十四卦

59 渙
Huàn
DISPERSING
Dispersion, dispersal, dissolution

☴ 巽 Xùn = 風 Wind
GROUND, the gentle, gentle penetration, entrance, flexibility
　　Element: 木 Wood 1

☵ 坎 Kǎn = 水 Water
GORGE, The Abysmal, dangerous, rapid rivers, the abyss, the moon
　　Element: 水 Water 5

60 ▦ The 64 Hexagrams 六十四卦

 **60 節 Jié**
ARTICULATING
Limitation, discipline, moderation

☵ 坎 Kǎn = 水 Water
GORGE, The Abysmal, dangerous, rapid rivers, the abyss, the moon
　　Element: 水 Water 5

☱ 兌 Duì = 澤 Marsh, Swamp
OPEN, the joyous, pleasure, satisfaction, stagnation
　　Element: 金 Metal 4

61 ▤ The 64 Hexagrams 六十四卦

61 中孚
Zhōng Fú
Inner Truth, confirming, central return

☴ Xùn = 風 Wind
GROUND, the gentle, gentle penetration, entrance, flexibility
　　Element: 木 Wood 1

☱ Duì = 澤 Marsh, Swamp
OPEN, the joyous, pleasure, satisfaction, stagnation
　　Element: 金 Metal 4

62 ▤ The 64 Hexagrams 六十四卦

62 小過
Xiǎo Guò
SMALL EXCEEDING
Small Preponderance, preponderance of the small, small surpassing

☳ Zhèn = 雷 Thunder
SHAKE, The Arousing, inciting, movement, initiative, revolution
　　Element: 木 Wood 1

☶ Gèn = 山 Mountain
BOUND, stillness, keeping still, resting, unable to be moved
　　Element: 土 Earth 3

63 ▤ The 64 Hexagrams 六十四卦

63 既濟
Jì Jì
ALREADY FORDING
After Completion, already completed, already done

☵ Kǎn = 水 Water
GORGE, The Abysmal, dangerous, rapid rivers, the abyss, the moon
　　Element: 水 Water 5

☲ Lí = 火 Fire
RADIANCE, The Clinging, rapid movement, clarity, adaptable, the sun
　　Element: 火 Fire 2

64 ▤ The 64 Hexagrams 六十四卦

64 未濟
Wèi Jì
NOT YET FORDING
Before Completion, not yet completed

☲ Lí = 火 Fire
RADIANCE, The Clinging, rapid movement, clarity, adaptable, the sun
　　Element: 火 Fire 2

☵ Kǎn = 水 Water
GORGE, The Abysmal, dangerous, rapid rivers, the abyss, the moon
　　Element: 水 Water 5

The Nine Northern Dipper Stars 北斗九星 (The Nine Great Emperors 九皇大帝)

The Nine Northern Dipper Stars (北斗九星 Běi Dǒu Jiǔ Xīng) or the Nine Great Emperors (九皇大帝 Jiǔ Huáng Dà Dì) are the sons of Dǒu Mǔ 斗母, the Goddess of the North Star and Queen of Heaven. She is number 17 in the Tarot Major Arcana known in Western countries as The Star. Her Chinese names include 斗母元君 Dǒu Mǔ Yuán Jūn and 北斗母九真圣德天后 Běi Dǒu Jiǔ Zhēn Shèng Dé Tiān Hòu. The Nine Great Emperor Stars live in an area in the sky which in ancient Chinese astronomy was known as the Purple Forbidden City (紫微垣 Zǐ Wēi Yuán) and Beijing's Forbidden City is named after it. The 8 Trigrams (Bā Guà) are related to eight of the nine stars of the Flying Star Fēng Shuǐ. The fifth star has no Bā Guà trigram but is related to Centre and Earth. Each of the other stars is related to one of the 8 directions.

3 ⭐ Nine Northern Dipper Stars
北斗九星 (九皇大帝)

γ UMa
Gamma Ursae Majoris **1** 木 Wood

东 E
Conste-llations 42

三碧
3. Jade

禄存星
Lù Cún

天玑宫

Zhèn 震 ☳

Lasting Prosperity Star of True Man (Phecda : Thigh of the Bear)

4 ⭐ Nine Northern Dipper Stars
北斗九星 (九皇大帝)

δ UMa
Delta Ursae Majoris **1** 木 Wood

东南 SE
Conste-llations 42

四绿
4. Green

文曲星
Wén Qǔ

天权宫

Xùn 巽 ☴

Civil Star of Mystery and Darkness (Megrez : Root of the Tail of the Bear)

5 ⭐ Nine Northern Dipper Stars
北斗九星 (九皇大帝)

ε UMa
Epsilon Ursae Majoris **3** 土 Earth

中 Centre
Conste-llations 42

五黄
5. Yellow

廉贞星
Lián Zhēn

Lunar Mansion 25

玉衡宫

Earthly Branch **7** 午马 Horse

Sincere Star of Honesty and Chastity (Alioth : Black Horse)

6 ⭐ Nine Northern Dipper Stars
北斗九星 (九皇大帝)

ζ UMa
Zeta Ursae Majoris **4** 金 Metal

西北 NW
Conste-llations 42

六白
6. White

武曲星
Wǔ Qǔ

开阳宫

Qián 乾 ☰

Military Star of the North Pole (Mizar : Girdle)

7 ✪	Nine Northern Dipper Stars 北斗九星 (九皇大帝)
	η UMa Eta Ursae Majoris **4** 金 Metal

西 W

Constellations 42

七赤 7. Red   破军星 Pò Jūn

摇光宫

Duì 兑 ☱

Army Defeating Star of Heaven's Gate (Alkaid : End of the Tail of the Bear)

---

8 ✪	Nine Northern Dipper Stars 北斗九星 (九皇大帝)
Chemical Elements 61	α Lyr Alpha Lyrae **3** 土 Earth

东北 NE

Constellations 69, 74

Planets & Moons 7/5

八白   左辅星
8. White Left Assistant

Gōu Chén
勾陈天皇大帝

Gèn 艮 ☶

Great Heavenly Emperor of the Curved Array (Vega: from Arabic *an-nasr al-waqi* "The falling Eagle", Latin: *Vultur Cadens*)

---

9 ✪	Nine Northern Dipper Stars 北斗九星 (九皇大帝)
	α UMi Alpha Ursae Minoris **2** 火 Fire

南 S

Constellations 54

九紫   右弼星
9. Purple Right Assistant

Zǐ Wēi
北极紫微大帝

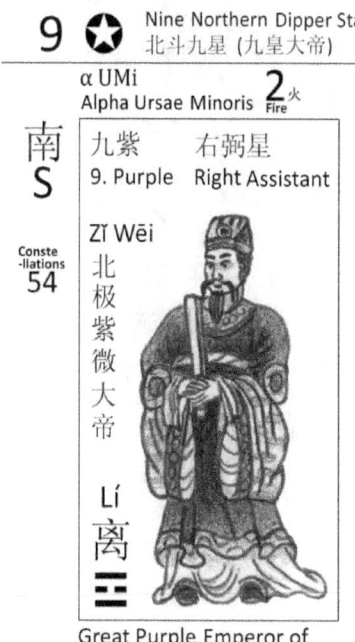

Lí 离 ☲

Great Purple Emperor of the North Pole (Polaris : The Pole Star)

The 6 Southern Dipper Stars 南斗六星

The Six Southern Dipper Stars 南斗六星 (Nán Dǒu Liù Xīng) are all within the constellation Sagittarius and five of them form what astronomers call the upside-down Little Milk Dipper. The sixth star is Mu Sagittarii which is actually a star system having the traditional name Polis, a Coptic word meaning "foal". The arrow of the bow held by Sagittarius (the Archer), together with the Little Milk Dipper, form "the Teapot".

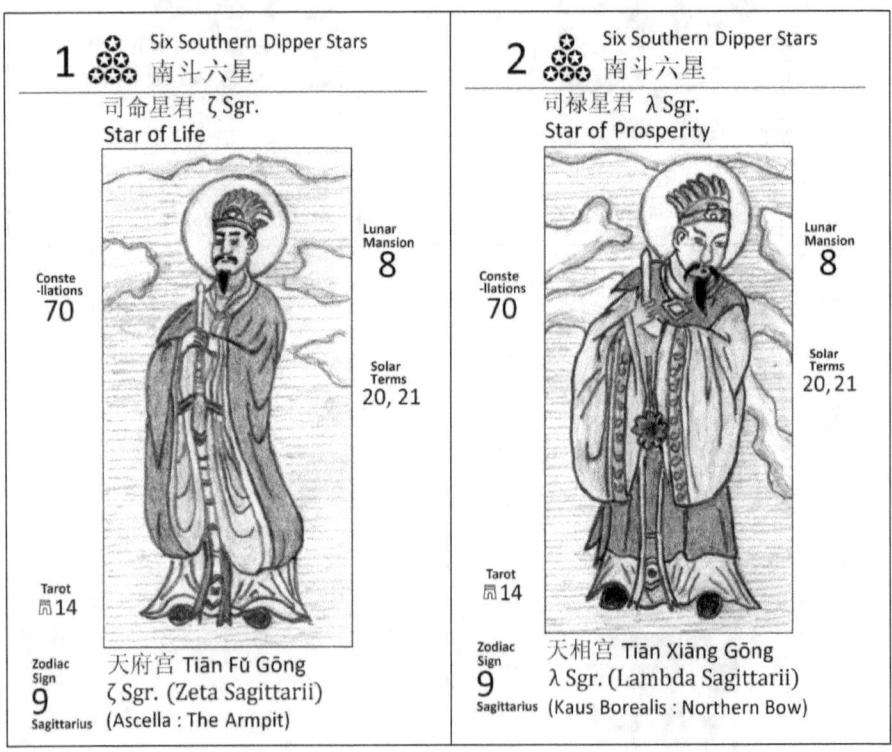

The Nine Sons of the Dragon and 3 more
龙生九子

The Ming Dynasty authors Lù Róng 陆容 (1436-1494), Yáng Shèn 杨慎 (1488-1559) and Xiè Zhào-zhè 谢肇浙 (1567-1624) claimed that the dragon had nine sons and this concept later entered popular Chinese culture. However, their lists differed somewhat from each other and some dragons had alternative names causing some confusion. Nine is the largest single digit and is often related to dragons. Nine dragon walls in Beijing's Forbidden City and Beihai Park are examples and the name Kowloon (九龙 Jiǔ Lóng) in Hong Kong means "nine dragons". Nine may also indicate a large number, sometimes more than just nine. Images of 12 sons of the dragon are displayed here but the money attracting Pí Xiū 貔貅 and Lóng Guī 龙龟 are also regarded by some authorities as sons of the dragon, making a more or less maximum number of 14. These drawings are largely based on statues of the Nine Sons of the Dragon on Shenyang's Huī Shān 辉山. Bà Xià 霸下 (赑屃 Bì Xi) is not to be confused with Bā Xià 趴蝮 (蚣蝮 Gōng Fù).

Nine Sons of the Dragon and more
龙生九子
囚牛 Qiú Niú
Music Dragon
1 Wood

The head of Qiú Niú is carved on stringed instruments.

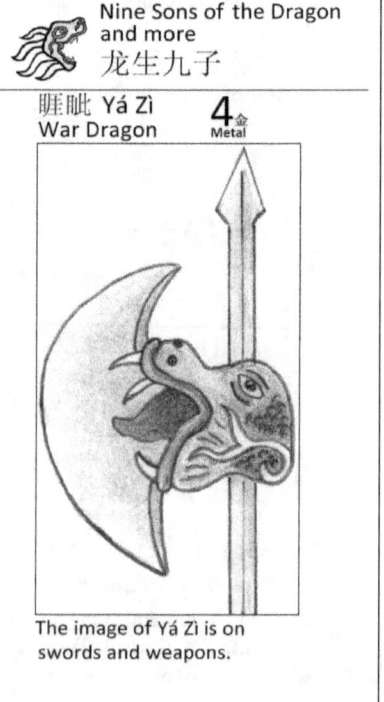

Nine Sons of the Dragon and more
龙生九子
睚眦 Yá Zì
War Dragon
4 Metal

The image of Yá Zì is on swords and weapons.

3 Nine Sons of the Dragon and more
龙生九子

狴犴 Bì Àn
Dragon Dog

2 火 Fire

Lunar Mansion **22**

Tarot 网 11

Bì Àn is the Dragon Dog on the doors and walls of law courts and prisons. He symbolizes justice.

4 Nine Sons of the Dragon and more
龙生九子

狻猊 Suān Ní
Dragon Lion

2 火 Fire

Suān Ní is the Dragon Lion on incense burners in temples.

5 Nine Sons of the Dragon and more
龙生九子

饕餮 Tāo Tiè
Gourmet Dragon

Tāo Tiè attends banquets and is symbolized by his own Tāo Tiè patterns on crockery and tableware.

6 Nine Sons of the Dragon and more
龙生九子

椒图 Jiāo Tú
Dragon Door Guard

3 土 Earth

Jiāo Tú is the Dragon on door knockers, door stoppers and door posts.

7 — **Nine Sons of the Dragon and more** 龙生九子 贔屭 **Bì Xī** — Stone Dragon Tortoise **5** 水 Water  Tarot ♡24 Bì Xī is the Stone Dragon Tortoise which supports inscribed stone stellai. It is also called Bà Xià 霸下	**8** — **Nine Sons of the Dragon and more** 龙生九子 螭吻 **Chī Wěn** — Rooftop Dragon **5** 水 Water 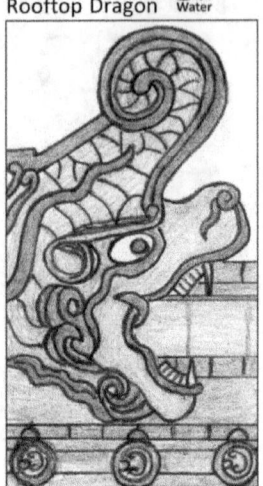 Chī Wěn is the Dragon on rooftops which sometimes has a fish tail.
9 — **Nine Sons of the Dragon and more** 龙生九子 蒲牢 **Pú Láo** — Bell Dragon 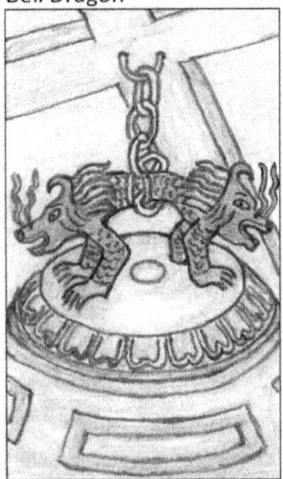 Pú Láo is the Dragon attached to the top of bells in bell towers	**10** — **Nine Sons of the Dragon and more** 龙生九子 嘲风 **Cháo Fēng** — Roof Corner Dragon  Cháo Fēng is the Dragon behind the Immortal riding a Phoenix 仙人骑凤 on the corners of roofs

| 11 | Nine Sons of the Dragon and more 龙生九子 | 12 | Nine Sons of the Dragon and more 龙生九子 |

负屃 Fù Xī
Literature Dragon

趴蝮 Bā Xià 　5水 Water
Water Dragon

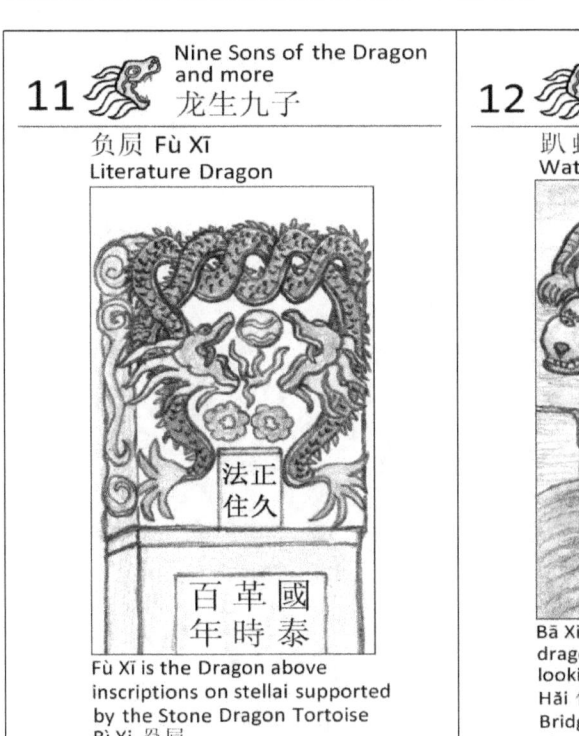

Fù Xī is the Dragon above inscriptions on stellai supported by the Stone Dragon Tortoise Bì Xī 赑屃

Bā Xià (趴蝮 or 蚣蝮) is a dragon that lies on its stomach looking at the waters of Shí Chà Hǎi 什刹海 near Hòu Mén Bridge 后门桥 in Beijing. It is also known as Gōng Fù 蚣蝮

The 112 Chemical Elements 化学元素

These are artistic representations of 112 Chemical Elements (化学元素 Huà Xué Yuán Sù). Each is listed in order of its atomic number or number of protons in its nucleus. Images representing further elements such as 114 Flerovium and 116 Livermorium will be produced later. Many element names are derived from Greek mythology with images of gods, goddesses and ancient heroes employed to represent them. Others are represented by the products in which they commonly occur, such as fluoride toothpaste (Fluorine) or by the places they were found or named after, such as Ytterby Village Mine (Terbium) or coins of the Parisii people of ancient Lutetia now called Paris (Lutetium). Portraits of discoverers, such as the Finnish chemist Johan Gadolin (1760-1853) are also employed. Praseodymium meaning "green twin" is represented by nature goddess Artemis, twin brother of Apollo, walking beside a deer in the forest. The image for Neodymium "new twin" shows her brother Apollo.

1 The Chemical Elements 化学元素

H Chemical Elements 1 | Air 空气 | 1 Wood 木

氢 Hydrogen
Qīng

Hydrogen filled airship Hindenburg explodes in 1937
1937 年氢气弥漫地兴登堡号飞艇爆炸

2 The Chemical Elements 化学元素

He Chemical Elements 2 | Air 空气 | 1 Wood 木 | Day 7 Sun

氦 Helium
Hài

Shang Suns 3
Planets & Moons 0, 6/66
Asteroids & Moons, Comets 1, 10, 93
风 19

Helium was named after Greek Sun God Helios
以古希腊太阳神赫利俄斯命名。

3 The Chemical Elements 化学元素

Li Chemical Elements 3 | 4 金 Metal

锂 Lithium
Lǐ

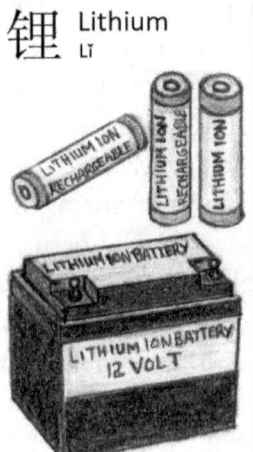

Rechargeable Lithium batteries
可充电的锂电池

4 The Chemical Elements 化学元素

Be Chemical Elements 4 | 4 金 Metal

铍 Beryllium
Pí

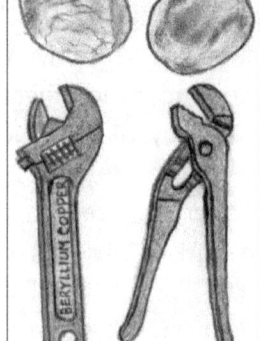

Beryl gem stones and beryllium-copper tools
绿宝石和铍铜合金工具

5  The Chemical Elements 化学元素	6 The Chemical Elements 化学元素
Chemical Elements 5 **B** 3 上 Earth 硼 Boron Péng  Boron used in making bakeware and laboratory glassware 硼用来制作烘焙用具和实验室玻璃器皿。	Chemical Elements 6 **C** 3 上 Earth 碳 Carbon Tàn 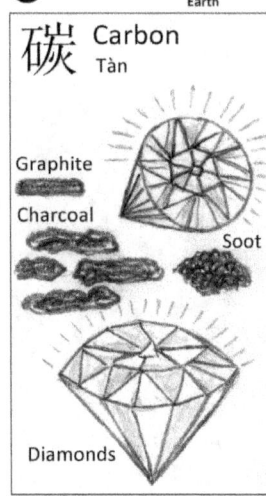 Graphite Charcoal Soot Diamonds Diamonds, charcoal, soot and graphite are forms of carbon. 碳元素存在于钻石，木炭，煤烟和石墨中。
7 The Chemical Elements 化学元素	8 The Chemical Elements 化学元素
Chemical Elements 7 **N** Air 空气 1 木 Wood 氮 Nitrogen Dàn 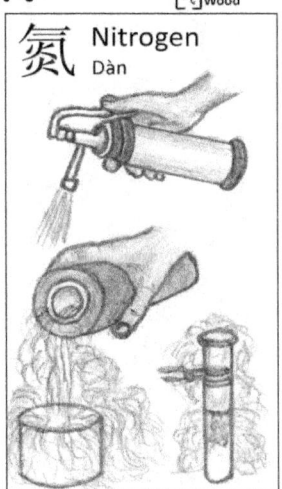 Liquid nitrogen 液态氮	Chemical Elements 8 **O** Air 空气 1 木 Wood 氧 Oxygen Yǎng  Oxygen cylinders used in climbing Mount Everest (Zhū Mù Lǎng Mǎ) 攀登珠穆朗玛峰使用氧气瓶。

9	The Chemical Elements 化学元素

| Chemical Elements 9 | **F** | | Air 空气 | 1 木 Wood |

氟 **Fluorine**
Fú

Fluoride toothpaste
含氟牙膏

10	The Chemical Elements 化学元素

| Chemical Elements 10 | **Ne** | | Air 空气 | 1 木 Wood |

氖 **Neon**
Nǎi

Neon sign
霓虹灯广告牌

---

11 The Chemical Elements 化学元素

| Chemical Elements 11 | **Na** | | 4 金 Metal |

钠 **Sodium**
Nà

Salt is a compound of sodium and chlorine.
盐是钠和氯的化合物。

---

12 The Chemical Elements 化学元素

| Chemical Elements 12 | **Mg** | ☾+○+ | 4 金 Metal |

镁 **Magnesium**
Měi

Conste-llations
16, 80, 86

Planets & Moons
6/6, 6/39, 7/20, 7/41, 9/3

Asteroids & Moons, Comets
7, 80, 98,111

Tarot
♣ 27
♣♣◇♡
12

Zodiac Sign
2
Taurus

Lunar Mansion
9, 25

Named after Magnesia. Coin of Magnesia c. 150 BC
约公元前150年古马格尼西亚地区钱币，镁元素以此地区命名。

Earthly Branch
2, 7

13 ⚛ The Chemical Elements 化学元素

Chemical Elements 13 **Al** **4** 金 Metal

铝 **Aluminium** Lǚ

Planets & Moons **8/9**

Asteroids & Moons, Comets **8, 14, 72**

Aluminium statue of Anteros, Piccadilly Circus, London, cast in 1893 安忒洛斯铝制雕像，铸造于1893年，位于伦敦皮卡迪利广场。

14 ⚛ The Chemical Elements 化学元素

Chemical Elements 14 **Si** **3** 土 Earth

硅 **Silicon** Guī

Silicate minerals are in whiteware ceramics, porcelain and concrete. 硅酸盐矿物存在于陶瓷器皿和水泥中。

15 ⚛ The Chemical Elements 化学元素

Chemical Elements 15 **P** ♃ △ ⌒ **3** 土 Earth **Day 5** Fri

磷 **Phosphorus** Lín

Planets & Moons **2**

Phosphorus (Eosphoros) the Morning Star, God of Dawn 以启明星之神福斯福洛斯（又名厄俄斯福洛斯）命名。

16 ⚛ The Chemical Elements 化学元素

Chemical Elements 16 **S** ⛢ ⛢ ▯ ⚵ **3** 土 Earth

硫 **Sulfur** Liú

Yuan Dynasty Mongol soldier lights a rocket

14th Century Ming Dynasty proto-cannon

Sulfur is a component of gunpowder. 硫是火药的一个成分。

17 The Chemical Elements
化学元素

Chemical Elements 17 **Cl** Air 空气 1 木 Wood

氯 **Chlorine**
Lǜ

Asteroids & Moons, Comets
17, 54

Chloris Greek nymph of flowers 以希腊神话的花神克洛里斯命名。

18 The Chemical Elements
化学元素

Chemical Elements 18 **Ar** Air 空气 1 木 Wood

氩 **Argon**
Yà

Argon gas used in fire extinguishers
氩气使用于灭火器中。

19 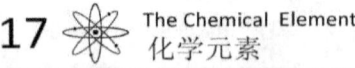 The Chemical Elements
化学元素

Chemical Elements 19 **K** ♄♃♃ 4 金 Metal

钾 **Potassium**
Jiǎ

Adding potassium based fertilizer to the soil
在土壤中添加钾肥。

20 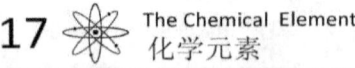 The Chemical Elements
化学元素

Chemical Elements 20 **Ca** 4 金 Metal

钙 **Calcium**
Gài

Calcium rich milk and cheese
高钙牛奶和奶酪。

21 ⚛ The Chemical Elements 化学元素	22 ⚛ The Chemical Elements 化学元素
Chemical Elements 21 — **Sc** — 4 金 Metal	Chemical Elements 22 — **Ti** — 4 金 Metal
钪 Scandium Kàng 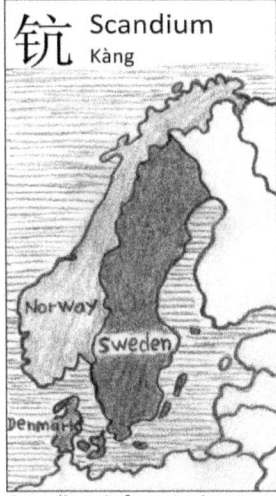 Scandium is from Latin Scandia (Scandinavia) 以斯堪的纳维亚的拉丁名字命名。	钛 Titanium Tài Planets & Moons 7/4, 7/22  The Titan Atlas holds up the sky. Titans imprisoned in the underworld 提坦巨神阿特拉斯托举着天空，其它的提坦神被囚禁在地狱。

23 ⚛ The Chemical Elements 化学元素	24 ⚛ The Chemical Elements 化学元素
Chemical Elements 23 — **V** — 4 金 Metal — Day 5 Fri	Chemical Elements 24 — **Cr** — 4 金 Metal
钒 Vanadium Fán Asteroids & Moons, Comets 47 Tarot ♠20 Vanadis (Freya) rides in a chariot pulled by cats 凡娜迪丝（弗蕾亚）驾着猫拉的战车。	铬 Chromium Gè  Chrome grill on cars in the 1950s 20世纪50年代汽车上的镀铬格栅

| 25 | ⚛ | The Chemical Elements 化学元素 |

Chemical Elements 25

Mn 4 金 Metal

锰 Manganese Měng

Conste-llations 80, 86

Planets & Moons 6/39, 7/20, 9/3

Asteroids & Moons, Comets 98, 111

Tarot ♠12 ♣12 ♢12 ♡12

Lunar Mansion 25

Earthly Branch 7 马 Horse

Manganese based pigments in Lascaux cave paintings
法国拉斯科洞窟壁画的颜料基于锰元素。

| 26 | ⚛ | The Chemical Elements 化学元素 |

Chemical Elements 26

Fe ♂ ⚔ 4 金 Metal

铁 Iron Tiě

Ming Dynasty blast furnace for smelting iron
明朝用鼓风炉炼铁。

| 27 | ⚛ | The Chemical Elements 化学元素 |

Chemical Elements 27

Co 4 金 Metal

钴 Cobalt Gǔ

Planets & Moons 8/16

Mischievous German Goblin Kobold
顽皮的德国地精科博尔德。

| 28 | ⚛ | The Chemical Elements 化学元素 |

Chemical Elements 28

Ni 4 金 Metal

镍 Nickel Niè

Planets & Moons 8/16

Nickel, malevolent goblin of German mines
德国矿山上的恶毒的地精。

Astronomical Pencil Drawings - John Oxenham Goodman

29 The Chemical Elements 化学元素

Chemical Elements 29 Cu ⚛ ☿ ♁ 4 金 Metal

铜 Copper
Tóng

Eighty tons of copper in Statue of Liberty
自由女神由 80 吨铜铸成。

30 The Chemical Elements 化学元素

Chemical Elements 30 Zn ⚛ ⬡ 4 金 Metal

锌 Zinc
Xīn

Galvanized roofing, galvanized angle iron
镀锌屋顶和角铁。

31 The Chemical Elements 化学元素

Chemical Elements 31 Ga 4 金 Metal

镓 Gallium
Jiǎ

Lunar Mansion 25

Conste-llations 80, 86

Planets & Moons 6/39, 7/20, 9/3

Asteroids & Moons, Comets 98, 111

Tarot
♠12
♣12
♦12
♥12

Gallium is named after Gaul. Vercingetorix chieftain of the tribes of Gaul 以高卢命名，高卢人首领韦辛格托里克斯。

Earthly Branch 7 午马 Horse

32 The Chemical Elements 化学元素

Chemical Elements 32 Ge 3 土 Earth

锗 Germanium
Zhě

Conste-llations 69, 74

Planets & Moons 6, 6/7, 7/5

Asteroids & Moons, Comets 49, 76

German 10 Mark commemorative coin

Germanium is named after Germany.
以德国命名

33 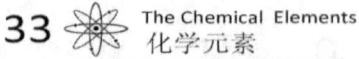 The Chemical Elements 化学元素

Chemical Elements 33

As ⚷ ♀ ♄ **3** 土 Earth

砷 Arsenic Shēn

Albertus Magnus (1193-1280) German bishop, philosopher, astrologer and scientist was first to isolate Arsenic.
艾尔伯图斯·麦格努斯是德国哲学家，主教，占星家及科学家，率先分离砷元素。

34 The Chemical Elements 化学元素

Chemical Elements 34

Se **3** 土 Earth

硒 Selenium Xī

Planets & Moons **3/1**

Tarot 2,18

Greek Moon Goddess Selene
塞勒涅—古希腊月亮女神

35 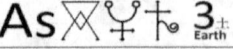 The Chemical Elements 化学元素

Chemical Elements 35

Br **5** 水 Water

溴 Bromine Xiù

Tyrian purple dye is a compound of bromine. Byzantine Empress Theodora I wears purple robes.
骨螺紫染料是一种溴化合物，拜占庭帝国狄奥多拉皇后穿着骨螺紫长袍。

36 The Chemical Elements 化学元素

Chemical Elements 36

Kr Air 空气 **1** 木 Wood

氪 Krypton Kè

Krypton used in fluorescent lamps and photography
用于荧光灯和摄影灯。

37 The Chemical Elements 化学元素 Chemical Elements 37 **Rb** 4 金 Metal 铷 Rubidium Rú 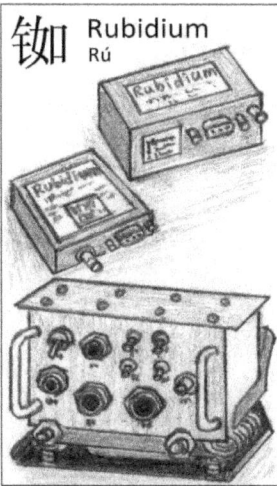 GPS rubidium clocks. Rubidium frequency standard. Miniature rubidium oscillator. GPS 铷钟。	**38** The Chemical Elements 化学元素 Chemical Elements 38 **Sr** 4 金 Metal 锶 Strontium Sī  Strontian, Scotland (Sròn an t-Sithein) and mythical Aos Sí people of the mounds (the Sidhe or Sith). 名字来源于苏格兰村庄斯特郎廷，有一群仙人居住在那里的山上。
39 The Chemical Elements 化学元素 Chemical Elements 39 **Y** 4 金 Metal 钇 Yttrium Yǐ  Vaxholm Coat of Arms Yttrium was found in mine at Ytterby Village, Vaxholm, Sweden 以发现地伊特比矿命名，位于瑞典的瓦克斯霍尔姆城。	**40** The Chemical Elements 化学元素 Chemical Elements 40 **Zr** 4 金 Metal 锆 Zirconium Gào  Zirconium dioxide used in laboratory crucibles and grinding wheels 二氧化锆用于制作实验室坩埚和砂轮。

41 The Chemical Elements 化学元素

Chemical Elements 41 **Nb** 4 Metal

铌 Niobium Ní

Asteroids & Moons, Comets 59

10 of Niobe's 12 children were killed by Apollo and Artemis.
尼奥比因自夸而被阿波罗和阿耳特弥斯杀死 10 个孩子

42 The Chemical Elements 化学元素

Chemical Elements 42 **Mo** 4 Metal Day 7 Sun

钼 Molybdenum Mù

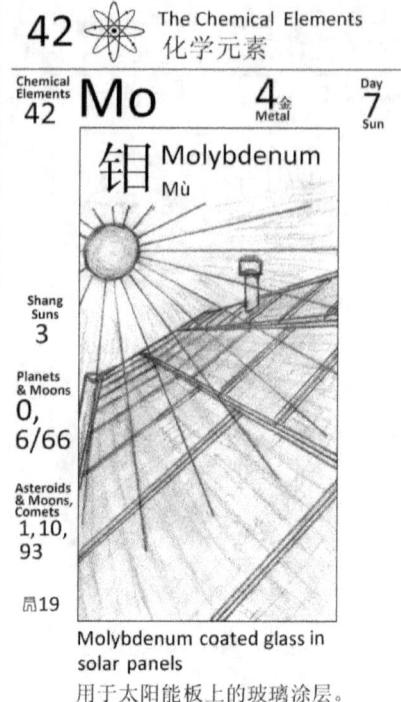

Shang Suns 3

Planets & Moons 0, 6/66

Asteroids & Moons, Comets 1, 10, 93

商19

Molybdenum coated glass in solar panels
用于太阳能板上的玻璃涂层。

43 The Chemical Elements 化学元素

Chemical Elements 43 **Tc** 4 Metal

锝 Technetium Dé

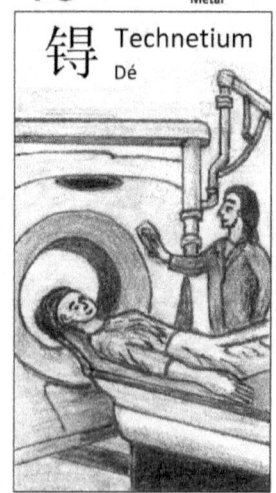

Technetium used as a radioactive tracer in medical tests 用于医学检验上的放射性示踪剂。

44 The Chemical Elements 化学元素

Chemical Elements 44 **Ru** 4 Metal

钌 Ruthenium Liǎo

Conste-llations 69, 74

Planets & Moons 6, 6/7, 7/5

Asteroids & Moons, Comets 49, 76

Russian Coat of Arms

Ruthenia, Latin name for Russia
俄罗斯的拉丁名字。

45 ⚛ The Chemical Elements 化学元素	
Chemical Elements 45	**Rh** 4 金 Metal

铑 Rhodium
Lǎo

The dying nymph Rodanthe was turned into a rose.
一个垂死的仙女罗丹特被变成了玫瑰。

46 ⚛ The Chemical Elements 化学元素	
Chemical Elements 46	**Pd** 4 金 Metal

钯 Palladium
Bǎ

Asteroids & Moons, Comets
58, 66

Cassandra clings to the Palladium for protection
卡珊德拉向守护神 Palladium 寻求保护。

47 ⚛ The Chemical Elements 化学元素	
Chemical Elements 47	**Ag** ⚷ ⚵ 4 金 Metal

银 Silver
Yín

Asteroids & Moons, Comets
58, 66

Silver tetradrachm of Athens. Athena and owl 450 BC
雅典四德拉克马银币的正面是雅典娜头像，反面是猫头鹰像，约前 450 年。

48 ⚛ The Chemical Elements 化学元素	
Chemical Elements 48	**Cd** 4 金 Metal

镉 Cadmium
Gé

Conste-llations
55

Lunar Mansion
1, 2

Cadmus slays a water dragon at the Castalian Spring
卡德摩斯在卡斯塔利亚泉边上杀死一条水龙。

Earthly Branch
5 辰龙
Dragon

49 ⚛ The Chemical Elements 化学元素	50 ⚛ The Chemical Elements 化学元素
Chemical Elements 49 **In** 4 金 Metal	Chemical Elements 50 **Sn** ⚻ 🏹 4 金 Metal

铟 **Indium**
Yīn

锡 **Tin**
Xī

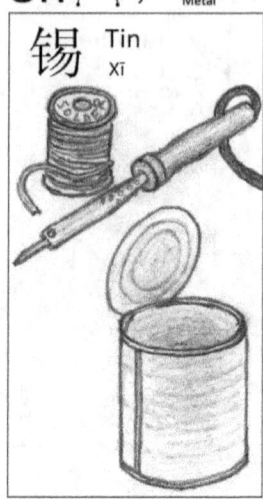

Indium used in crystal displays and touchscreens
用于液晶显示屏及触摸屏。

Tin is a component of lead-free solder; Tin plated cans
无铅焊锡中的一种化合物，用于焊接镀锡罐。

51 ⚛ The Chemical Elements 化学元素	52 ⚛ The Chemical Elements 化学元素
Chemical Elements 51 **Sb** ☤ ♁ ◇ 3 土 Earth	Chemical Elements 52 **Te** 3 土 Earth

锑 **Antimony**
Tī

碲 **Tellurium**
Dì

Planets & Moons
3

Antimony sulfide Sb₂S₃ (Stibnite) used as black eye paint in ancient Egypt
硫化锑用于古代埃及眼部化妆涂料。

Tellus, Roman Goddess of the Earth
特鲁斯是古罗马大地女神

125

53 ⚛ The Chemical Elements 化学元素

Chemical Elements 53   **I**   **3** 土 Earth

碘 Iodine
Diǎn

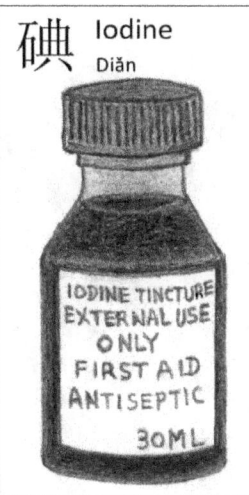

Iodine used as an antiseptic
用于消毒液。

54 ⚛ The Chemical Elements 化学元素

Chemical Elements 54   **Xe**   Air 空气 **1** 木 Wood

氙 Xenon
Xiān

Xenon used in house lights and car lights
用于室内灯和车灯。

55 ⚛ The Chemical Elements 化学元素

Chemical Elements 55   **Cs**   **4** 金 Metal

铯 Caesium
Sè

Caesium formate-based high pressure oil drilling
基于甲酸铯的高压石油钻井液。

56 ⚛ The Chemical Elements 化学元素

Chemical Elements 56   **Ba**   **4** 金 Metal

钡 Barium
Bèi

Barium gives green colour to fireworks
钡在烟花中呈现出绿色。

57 The Chemical Elements 化学元素

Chemical Elements 57 **La**

镧 Lanthanum
Lán

Lanthanum oxide La_2O_2 used in telescope and camera lenses
氧化镧用于望远镜和照相机的镜头。

58 The Chemical Elements 化学元素

Chemical Elements 58 **Ce**

铈 Cerium
Shì

Planets & Moons 5

Ceres, Roman Grain Goddess
谷神星刻瑞斯—古罗马女神

59 The Chemical Elements 化学元素

Chemical Elements 59 **Pr**

镨 Praseodymium
Pǔ

Planets & Moons 3/1

Tarot
♦2, 18
♠21, 22

Praseodymium "Green Twin"; Artemis twin sister of Apollo, goddess of nature, animals, hunting and the moon
名称源于希腊两词语"绿色"和"成对"的结合语，阿尔特弥斯是阿波罗的孪妹，是自然女神，猎神和月亮女神。

60 The Chemical Elements 化学元素

Chemical Elements 60 **Nd** Day 7 Sun

钕 Neodymium
Nǚ

Shang Suns 3

Planets & Moons
0, 6/66

Asteroids & Moons, Comets
1, 10, 93

♦19

Neodymium "New Twin"; Apollo, twin brother of Artemis, god of music, prophecy, archery and the sun
名称源于希腊两词语"新"和"成对"的结合语，阿波罗是阿尔特弥斯的孪兄，是乐神，预言家，箭神和太阳神。

61 ⚛ The Chemical Elements 化学元素

Chemical Elements 61
Pm

钷 Promethium
Pŏ

Conste-llations
69, 74

Planets & Moons
6, 6/7, 7/5

Asteroids & Moons, Comets
49, 76

Prometheus punished by Zeus
普罗米修斯被宙斯惩罚。

62 ⚛ The Chemical Elements 化学元素

Chemical Elements 62
Sm

钐 Samarium
Shān

Paul-Emile Lecoq de Boisbaudran (1838-1912) French chemist who discovered samarium
由法国化学家保罗・埃米尔・勒科克・德布瓦博德兰发现

63 ⚛ The Chemical Elements 化学元素

Chemical Elements 63
Eu

铕 Europium
Yŏu

Conste-llations
16

Asteroids & Moons, Comets
7, 80

Tarot
5, 27

Lunar Mansion
9

Earthly Branch
2
丑牛
Ox

Zodiac Sign
2
Taurus

Europa carried by a bull to Crete
欧罗巴被牛驮着来到克里特岛。

64 ⚛ The Chemical Elements 化学元素

Chemical Elements 64
Gd

钆 Gadolinium
Gá

Johan Gadolin (1760-1852) Finnish Chemist
以荷兰化学家约翰・加多林命名

65 ⚛ The Chemical Elements 化学元素 Chemical Elements 65 **Tb** 铽 **Terbium** Tè  Terbium found at Ytterby Village mine, Vaxholm, Sweden 以发现地伊特比矿命名，位于瑞典的瓦克斯霍尔姆城。	**66** ⚛ The Chemical Elements 化学元素 Chemical Elements 66 **Dy** 镝 **Dysprosium** Dī  Dysprosium oxide used in neutron absorbing control rods in nuclear reactors 氧化镝作为中子吸收剂，用于制作核反应堆的控制棒。
67 ⚛ The Chemical Elements 化学元素 Chemical Elements 67 **Ho** 钬 **Holmium** Huǒ  Stockholm Coat of Arms Holmia, Latin name of Stockholm, Sweden 以瑞典斯德哥尔摩的拉丁名字命名。	**68** ⚛ The Chemical Elements 化学元素 Chemical Elements 68 **Er** 铒 **Erbium** Ěr  Erbium discovered in quarry at Ytterby Village, Vaxholm, Sweden 以发现地伊特比采矿场命名，位于瑞典的瓦克斯霍尔姆城。

69 ⚛ The Chemical Elements 化学元素

Chemical Elements 69

Tm

铥 Thulium
Diū

Thule, Greek and Roman name of remote island in the North Atlantic 图勒是在古希腊和罗马时期北大西洋中的极北之地

---

70 ⚛ The Chemical Elements 化学元素

Chemical Elements 70

Yb

镱 Ytterbium
Yì

Vaxholm Fortress, Sweden near Ytterby Village Mine where Ytterbium was found 以发现地伊特比矿命名，瑞典的瓦克斯霍尔姆城堡垒。

---

71 ⚛ The Chemical Elements 化学元素

Chemical Elements 71

Lu

镥 Lutetium
Lǔ

Lunar Mansion 25

Conste-llations 80, 86

Planets & Moons 6/4, 6/39, 7/20, 9, 9/3

Asteroids & Moons, Comets 98, 111

Tarot ♣12 ♣12 ♢12 ♡12 ☐7

Lutetia, Latin name for Paris; coin of the Parisii people c. 50 BC 以巴黎的拉丁名字命名，巴黎西人的钱币，约公元前 50 年。

Earthly Branch 7 午马 Horse

---

72 ⚛ The Chemical Elements 化学元素

Chemical Elements 72

Hf

4 金 Metal

铪 Hafnium
Hā

Conste-llations 37, 40

Asteroids & Moons, Comets 50, 62, 70, 94

Tarot ☐8

Zodiac Sign 5 Leo

Copenhagen Coat of Arms

Hafnia, Latin name for Copenhagen 以哥本哈根的拉丁名字命名。

73	The Chemical Elements 化学元素

Chemical Elements 73 **Ta** **4 金 Metal**

钽 Tantalum
Tǎn

Asteroids & Moons, Comets 6

Tantalus condemned to eternal hunger and thirst
坦塔洛斯永远忍受饥渴的折磨。

74	The Chemical Elements 化学元素

Chemical Elements 74 **W** **4 金 Metal**

钨 Tungsten
Wū

Constellations 57
Planets & Moons 6/54, 7/29, 34, 39, 40, 46, 48, 57
Asteroids & Moons, Comets 97/1, 97/2

Tungsten (Swedish: heavy stone); Wolfram (German: wolf soot). Punishment of Sisyphus 瑞典语意为重石，德语意为狼灰。西绪福斯被判推着重石到山顶。

75	The Chemical Elements 化学元素

Chemical Elements 75 **Re** **4 金 Metal**

铼 Rhenium
Lái

Rhine River, Germany; Rheinstein Castle
以德国莱茵河命名，河西岸的莱茵石格堡。

76	The Chemical Elements 化学元素

Chemical Elements 76 **Os** **4 金 Metal**

锇 Osmium
É

Osmium used in fountain pen nibs and phonograph styli
用于钢笔尖和留声机唱针。

77 — The Chemical Elements 化学元素

Chemical Elements 77 **Ir** 4 金 Metal

铱 Iridium Yī

Asteroids & Moons, Comets 26

Iris, Goddess of the Rainbow
以彩虹女神爱莉丝命名。

78 — The Chemical Elements 化学元素

Chemical Elements 78 **Pt** ☽ ☉ ☿ 4 金 Metal

铂 Platinum Bó

Conste-llations 56

Tarot ◇6 ♣11

Zodiac Sign 7 Libra

American Platinum Eagle commemorative coin 2010
2010年美国发行的铂鹰纪念币。

79 — The Chemical Elements 化学元素

Chemical Elements 79 **Au** ☿ ☉ 4 金 Metal

金 Gold Jīn

Golden funerary mask of Tutankhamun (c. 1341-1323 BC)
图坦卡门的黄金面罩。

80 — The Chemical Elements 化学元素

Chemical Elements 80 **Hg** ☿ ♀ ♆ 4 金 Metal

Day 3 Wed

汞 Mercury Gǒng

Planets & Moons 1

Tarot ♣1

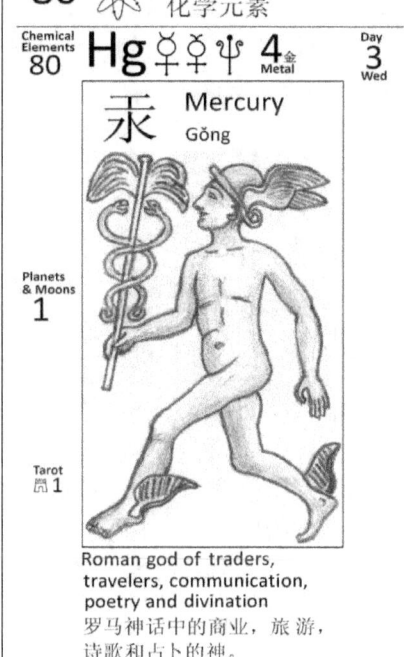

Roman god of traders, travelers, communication, poetry and divination
罗马神话中的商业，旅游，诗歌和占卜的神。

81	The Chemical Elements 化学元素

Chemical Elements 81 — Tl — 4 Metal

铊 Thallium Tā

Thallium: from Greek Thallos "a green shoot or twig"
源于希腊语，意为绿芽。

82	The Chemical Elements 化学元素

Chemical Elements 82 — Pb — 4 Metal

铅 Lead Qiān

Lead sinkers, lead bullets and lead wheel weights 用于铅锤，子弹及车轮上配重。

83	The Chemical Elements 化学元素

Chemical Elements 83 — Bi — 4 Metal

铋 Bismuth Bì

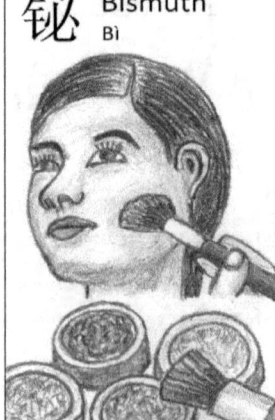

Bismuth oxychloride used in cosmetics
氯氧化铋用于化妆品。

84	The Chemical Elements 化学元素

Chemical Elements 84 — Po — 3 Earth

钋 Polonium Pō

Conste-llations 69, 74

Planets & Moons 6, 6/7, 7/5

Asteroids & Moons, Comets 49, 76

Coat of Arms of Poland

Polonia, Latin name for Poland
以波兰的拉丁名字命名。

85 The Chemical Elements
化学元素

Chemical Elements
85 **At** 3 土 Earth

砹 Astatine
Aì

Conste-llations
56

Planets & Moons
21/1

Tarot
◇6
🀫11

Zodiac Sign
7
Libra

Greek: Astatos "unstable"
源于希腊语，意为"不稳定的"。

86 The Chemical Elements
化学元素

Chemical Elements
86 **Rn** 1 木 Air 空气 Wood

氡 Radon
Dōng

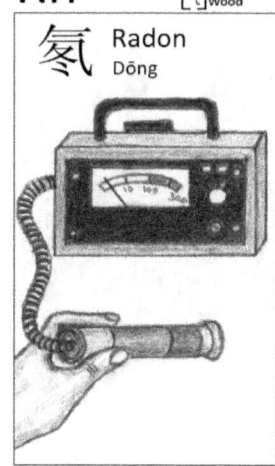

Geiger counter for detecting radon
用于测量放射性的盖革计数器。

87 The Chemical Elements
化学元素

Chemical Elements
87 **Fr** 4 金 Metal

钫 Francium
Fāng

Lunar Mansion
18

Tarot
🀫28

Francium named after France
以法国国名命名。

Earthly Branch
10
西鸡
Rooster

88 The Chemical Elements
化学元素

Chemical Elements
88 **Ra** 4 金 Metal

镭 Radium
Léi

Luminous radium paint on watches and clocks
用于钟表的发光表盘。

89 ⚛ The Chemical Elements 化学元素

Chemical Elements 89 **Ac** **Day 7 Sun**

锕 **Actinium** Ā

Shang Suns: 3
Planets & Moons: 0, 6/66
Asteroids & Moons, Comets: 1, 10, 93

冏 19

Greek: Aktinos "ray"; Actis, son of sun-god Helios, flees to Heliopolis in Egypt
阿克蒂斯，源于希腊语，意为"光线"，是太阳神赫利俄斯的儿子，逃到埃及的黑里欧波里斯。

90 ⚛ The Chemical Elements 化学元素

Chemical Elements 90 **Th** **Day 4 Thu**

钍 **Thorium** Tǔ

Planets & Moons: 6, 7/50

Thor, Norse God of Thunder
以北欧神话中的雷神托尔命名。

91 ⚛ The Chemical Elements 化学元素

Chemical Elements 91 **Pa** **Day 7 Sun**

镤 **Protactinium** Pú

Shang Suns: 3
Planets & Moons: 0, 6/66
Asteroids & Moons, Comets: 1, 10, 93

冏 19

From Greek: Proto Aktinos "first ray"; Actis, son of sun-god Helios, co-king of Rhodes
源于希腊语，意为"第一束光线"。阿克蒂斯是太阳神赫利俄斯的儿子，罗兹岛的统治者之一。

92 ⚛ The Chemical Elements 化学元素

Chemical Elements 92 **U**

铀 **Uranium** Yóu

Planets & Moons: 8
Asteroids & Moons, Comets: 25

Uranus (Heaven) holds the zodiac 以天王星尤拉诺斯命名，其手持十二宫星盘。

93 ⚛ The Chemical Elements 化学元素

Chemical Elements 93
Np

Conste-llation 80, 86

Sun 3/2

Planets & Moons 3/1, 6/4, 6/39, 7/20, 9, 9/3

Tarot
♠12
♣12
♢12
♡12
🀫 7

镎 **Neptunium** Ná

Lunar Mansion **25**

Neptune, Roman sea-god
以罗马海神尼普顿命名。

Earthly Branch **7** 午马 Horse

---

94 ⚛ The Chemical Elements 化学元素

Chemical Elements 94
Pu

Conste-llation 25, 29, 50

Planets & Moons 12

Asteroids & Moons, Comets 4

钚 **Plutonium** Bù

Lunar Mansion **16**

Pluto, giver of wealth, ruler of the underworld
以财神及地狱之王普路托命名。

Earthly Branch **11** 戌狗 Dog

---

95 ⚛ The Chemical Elements 化学元素

Chemical Elements 95
Am

镅 **Americium** Méi

The Americas named after Italian navigator Amerigo Vespucci (1454-1512)
以意大利航海家亚美利哥·韦斯普奇命名。

---

96 ⚛ The Chemical Elements 化学元素

Chemical Elements 96
Cm

锔 **Curium** Jú

Marie Curie (1867-1934)
Polish-born French physicist
Pierre Curie (1859-1906)
French physicist
居里夫人，波兰裔法国物理学家，皮埃尔·居里，法国物理学家。

97 ☢ The Chemical Elements 化学元素	98 ☢ The Chemical Elements 化学元素
Chemical Elements 97 **Bk** 铹 Berkelium Péi  University of California, Berkeley 加州大学伯克利分校。	**Chemical Elements 98** **Cf** 锎 Californium Kāi Conste-llation 42, 54 Planets & Moons 6/8, 6/24, 7/25 Asteroids & Moons, Comets 16 Bear flag raised on 14 June 1846. State flag of California since 1911. Ursus Californicus state animal of California since 1953. 以加利福尼亚命名。
99 ☢ The Chemical Elements 化学元素	100 ☢ The Chemical Elements 化学元素
Chemical Elements 99 **Es** 锿 Einsteinium Āi  Albert Einstein (1879-1955) German-born theoretical physicist 以德裔理论物理学家阿尔伯特·爱因斯坦命名。	**Chemical Elements 100** **Fm** 镄 Fermium Fèi Enrico Fermi (1901-1954) Italian physicist. Fermium discovered in debris of first hydrogen bomb detonated 1st November 1952 on Enewetak Atoll. 以意大利物理学家恩里科·费米命名，是在1952年位于埃内韦塔克环礁的第一次氢弹试验爆炸后的辐射落尘中发现的。

101 The Chemical Elements
化学元素

Chemical Elements 101 **Md**

钔 Mendelevium
Mén

Dmitri Ivanovich Mendeleev (1834-1907) Russian chemist
以俄国化学家季米特里·伊万诺维奇·门捷列夫命名。

---

102 The Chemical Elements
化学元素

Chemical Elements 102 **No**

锘 Nobelium
Nuò

Alfred Nobel (1833-1896) Swedish chemist, engineer and armaments manufacturer
以瑞典化学家，工程师及武器制造商阿尔弗雷德·诺贝尔命名。

---

103 The Chemical Elements
化学元素

Chemical Elements 103 **Lr**

铹 Lawrencium
Láo

Ernest Orlando Lawrence (1901-1958) American physicist 以美国物理学家欧内斯特·奥兰多·劳伦斯命名。

---

104 The Chemical Elements
化学元素

Chemical Elements 104 **Rf** 4 金 Metal

鑪 Rutherfordium
Lú

Baron Ernest Rutherford (1871-1937) New Zealand-born British physicist
以新西兰裔英国物理学家，男爵欧内斯特·卢瑟福命名。

105 ⚛ The Chemical Elements 化学元素

| Chemical Elements 105 | **Db** | 4 金 Metal |

钍 **Dubnium** Dù

Coat of Arms of Dubna, a city 200 kilometres from Moscow

以莫斯科附近的杜布那城命名。

106 ⚛ The Chemical Elements 化学元素

| Chemical Elements 106 | **Sg** | 4 金 Metal |

镭 **Seaborgium** Xǐ

Glenn Teodor Seaborg (1912-1999) American nuclear chemist

以美国核化学家格伦·西奥多·西博格命名。

107 ⚛ The Chemical Elements 化学元素

| Chemical Elements 107 | **Bh** | 4 金 Metal |

铍 **Bohrium** Bō

Niels Bohr (1885-1962) Danish physicist

以丹麦物理学家尼尔斯·波尔命名。

108 ⚛ The Chemical Elements 化学元素

| Chemical Elements 108 | **Hs** | 4 金 Metal |

镖 **Hassium** Hēi

Conste-ilations 37, 40

Asteroids & Moons, Comets 50, 62, 70, 94

Tarot 8

Coat of Arms of the German State of Hesse

Zodiac Sign 5 Leo

Hassia is the Latin name for Hesse or Hessen

以德国黑森州的拉丁名字命名。

109 ⚛ The Chemical Elements 化学元素

Chemical Elements 109 **Mt** **4 金 Metal**

镅 Meitnerium
Mài

Lise Meitner (1878-1968) Austrian and later Swedish nuclear physicist
以奥地利（后入瑞典藉）物理学家莉泽·迈特纳命名。

110 ⚛ The Chemical Elements 化学元素

Chemical Elements 110 **Ds** **4 金 Metal**

鿏 Darmstadtium
Dá

Conste-llations
37, 40

Asteroids & Moons, Comets
50, 62, 70, 94

Tarot
8

Zodiac Sign
5 Leo

Darmstadt Coat of Arms, Germany
以德国城市达姆施塔特命名。

111 ⚛ The Chemical Elements 化学元素

Chemical Elements 111 **Rg** **4 金 Metal**

铊 Roentgenium
Lún

Wilhelm Conrad Röntgen (1845-1923) German physicist
以德国物理学家威廉·康拉德·伦琴命名。

112 ⚛ The Chemical Elements 化学元素

Chemical Elements 112 **Cn** **4 金 Metal**

鎶 Copernicium
Gē

Nicolaus Copernicus (1473-1543) Polish mathematician and astronomer
以波兰数学家和天文学家尼古拉·哥白尼命名。

The 88 Constellations 八十八星座

The 88 constellations (八十八星座 Bā Shí Bā Xīng Zuò) are numbered in sequence around the celestial sphere beginning with Sculptor. Each is arranged in order of the right ascension of its midpoint starting from zero hours based on the constellation boundaries drawn up by Eugène Delporte in 1930 for the epoch B1875.0. Zodiacal constellations are circled. Chinese Lunar Mansions in the same part of the sky as some of the 88 constellations are underlined. Those Lunar Mansions not underlined such as 27 Snake resemble an animal of the 88 Constellations such as 63 Serpens. Similar animals are also found in the 12 Earthly Branches, the Planets and Moons, the Asteroids and Moons and the Tarot cards. An attempt has been made to assign Chinese elements to some of these constellations where they are obvious. For example 2 Pisces is assigned the element Water and 4 Phoenix the element Fire.

3 ☆ Constellations 八十八星座 **Andromeda** **5** 水 Water 仙女座 Constellations 3 Lunar Mansion **14, 15** 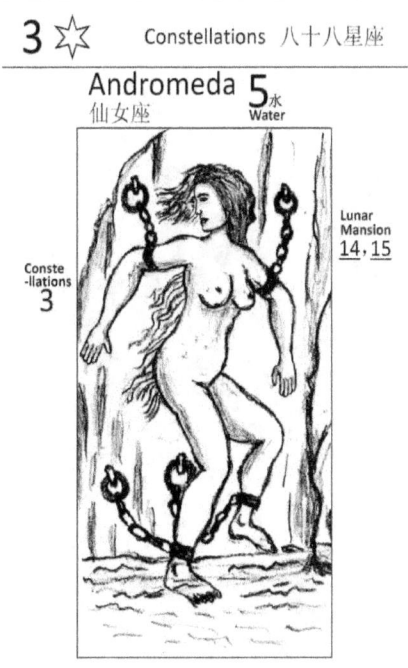 The Chained Princess	**4** ☆ Constellations 八十八星座 **Phoenix** **2** 火 Fire 凤凰座 Constellations 4  Tarot ◇27 ♠10 The Bird of Immortality
5 ☆ Constellations 八十八星座 **Cassiopeia** 仙后座 Constellations 5  The Seated Queen	**6** ☆ Constellations 八十八星座 **Cetus** 鲸鱼座 **5** 水 Water Constellations 6 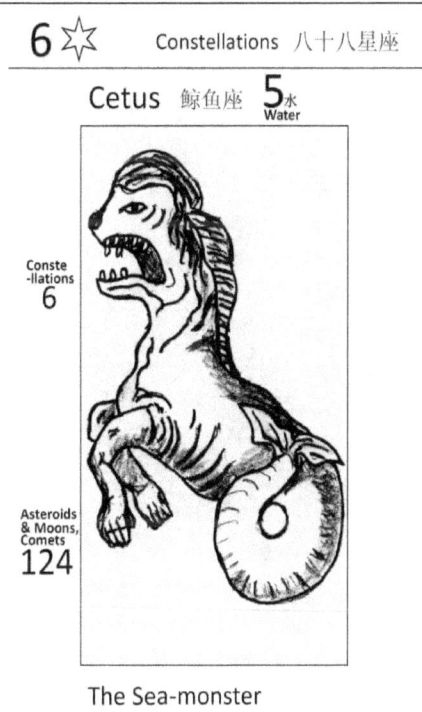 Asteroids & Moons, Comets **124** The Sea-monster

11 ☆ Constellations 八十八星座

Fornax 天炉座 **2** 火 Fire

Conste-llations 11

The Furnace

12 ☆ Constellations 八十八星座

Perseus 英仙座

Conste-llations 12

Asteroids & Moons, Comets 58/1, 58/2

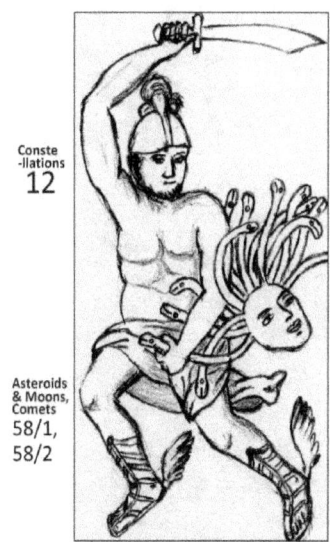

The Champion

13 ☆ Constellations 八十八星座

Horologium 时钟座

Conste-llations 13

The Clock

14 ☆ Constellations 八十八星座

Eridanus 波江座 **5** 水 Water

Conste-llations 14

The River Eridanus

15/1 ☆ Constellations 八十八星座

Reticulum 网罟座

Conste-llations
15/1

Reticulum Rhomboidalis
Rhomboidal Net

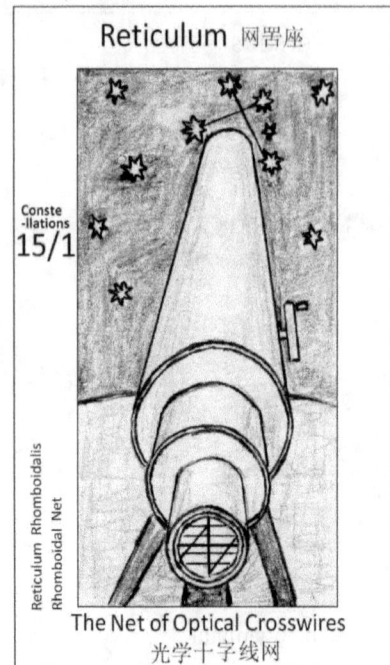

The Net of Optical Crosswires
光学十字线网

15/2 ☆ Constellations 八十八星座

Reticulum
网罟座

5 水
Water

Conste-llations
15/2

Tarot
♠19
♣1
♡21, 22, 23

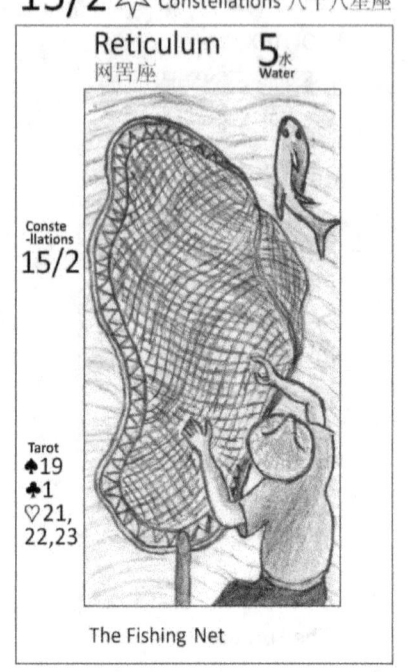

The Fishing Net

16 ☆ Constellations 八十八星座

Chemical Elements
12, 63

Taurus 金牛座

Conste-llations
(16)

Planets & Moons
6/6, 7/41

Asteroids & Moons, Comets
7, 80

Tarot
🀫 5, 27

Zodiac Sign
2
Taurus

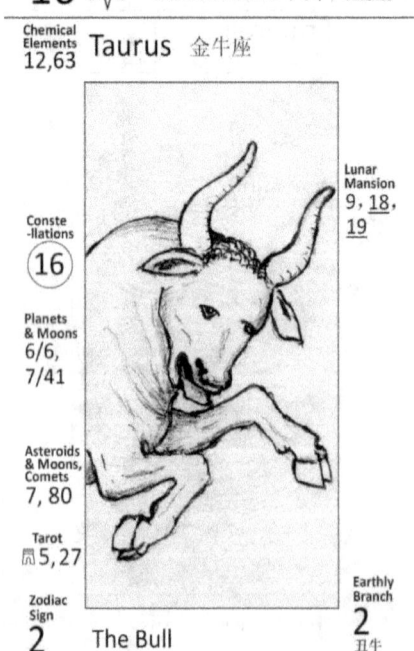

Lunar Mansion
9, **18**, **19**

Earthly Branch
2
丑牛
Ox

The Bull

17 ☆ Constellations 八十八星座

Caelum
雕具座

4 金
Metal

Conste-llations
17

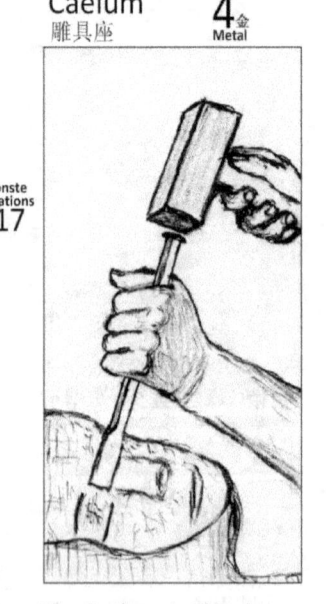

The Sculptor's Chisel

18 ☆ Constellations 八十八星座

Dorado 剑鱼座
5 水 Water

Constellations 18

The Swordfish

19 ☆ Constellations 八十八星座

Mensa 山案座
3 土 Earth

Tafelberg

Constellations 19

Cape Town Kaapstad

Table Mountain

20 ☆ Constellations 八十八星座

Lepus 天兔座

Constellations 20

Lunar Mansion **4**

Earthly Branch **4** 卯兔 Rabbit

The Hare

21 ☆ Constellations 八十八星座

Orion 猎户座

Constellations 21

Lunar Mansion **20, 21**

The Great Hunter

26/1 ☆ Constellations 八十八星座

Monoceros 麒麟座

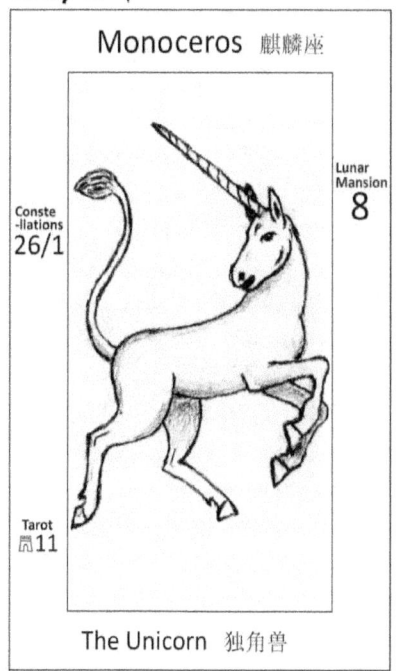

The Unicorn 独角兽

26/2 ☆ Constellations 八十八星座

Qí Lín

Equivalent of Monoceros in Chinese astronomy

27 ☆ Constellations 八十八星座

Gemini 双子座

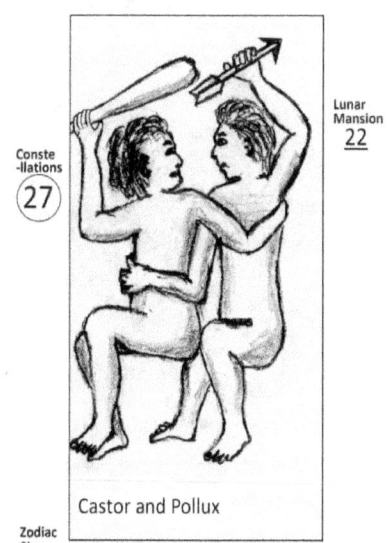

Castor and Pollux

The Twins

28 ☆ Constellations 八十八星座

Puppis 船尾座 5 水 Water

The Stern of the Argo

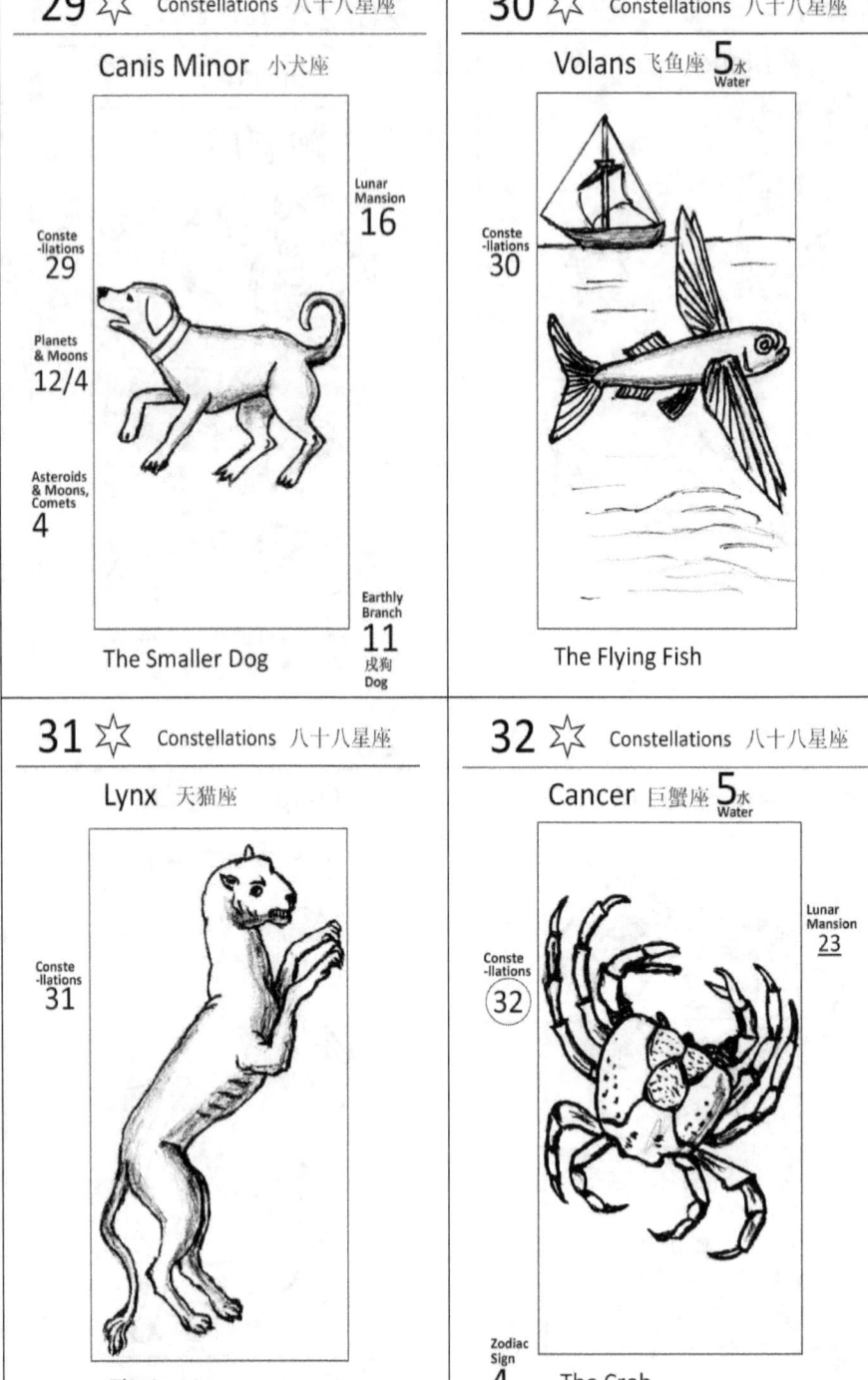

33 ☆ Constellations 八十八星座	34 ☆ Constellations 八十八星座
Carina 船底座 5 水 Water	Camelo Pardalis 鹿豹座
Constellations 33	Constellations 34
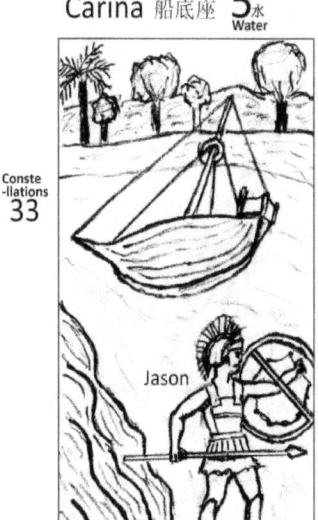	Tarot ♡9 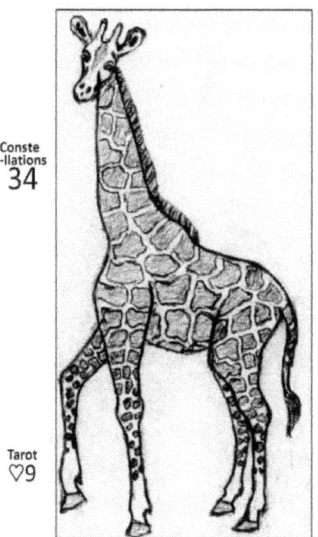
The Keel of the Argo	The Giraffe
35 ☆ Constellations 八十八星座	36 ☆ Constellations 八十八星座
Pyxis 罗盘座 5 水 Water	Vela 船帆座 5 水 Water
Constellations 35	Constellations 36
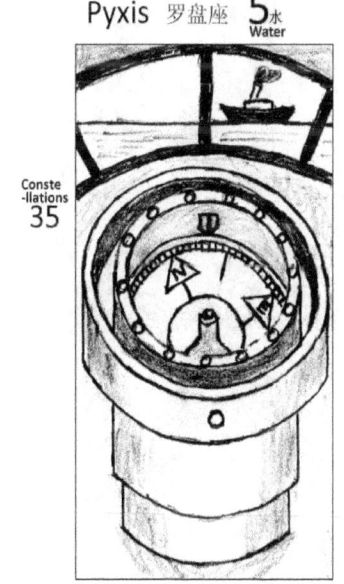	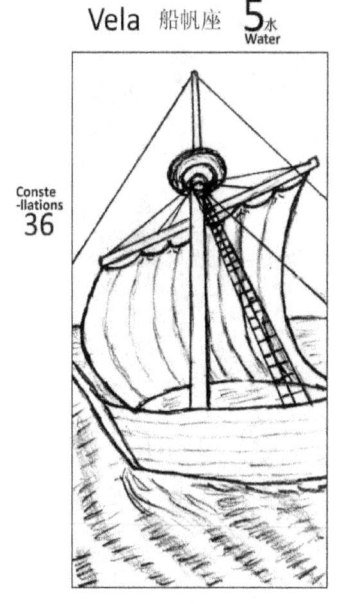
The Mariner's Compass	The Sails of the Argo

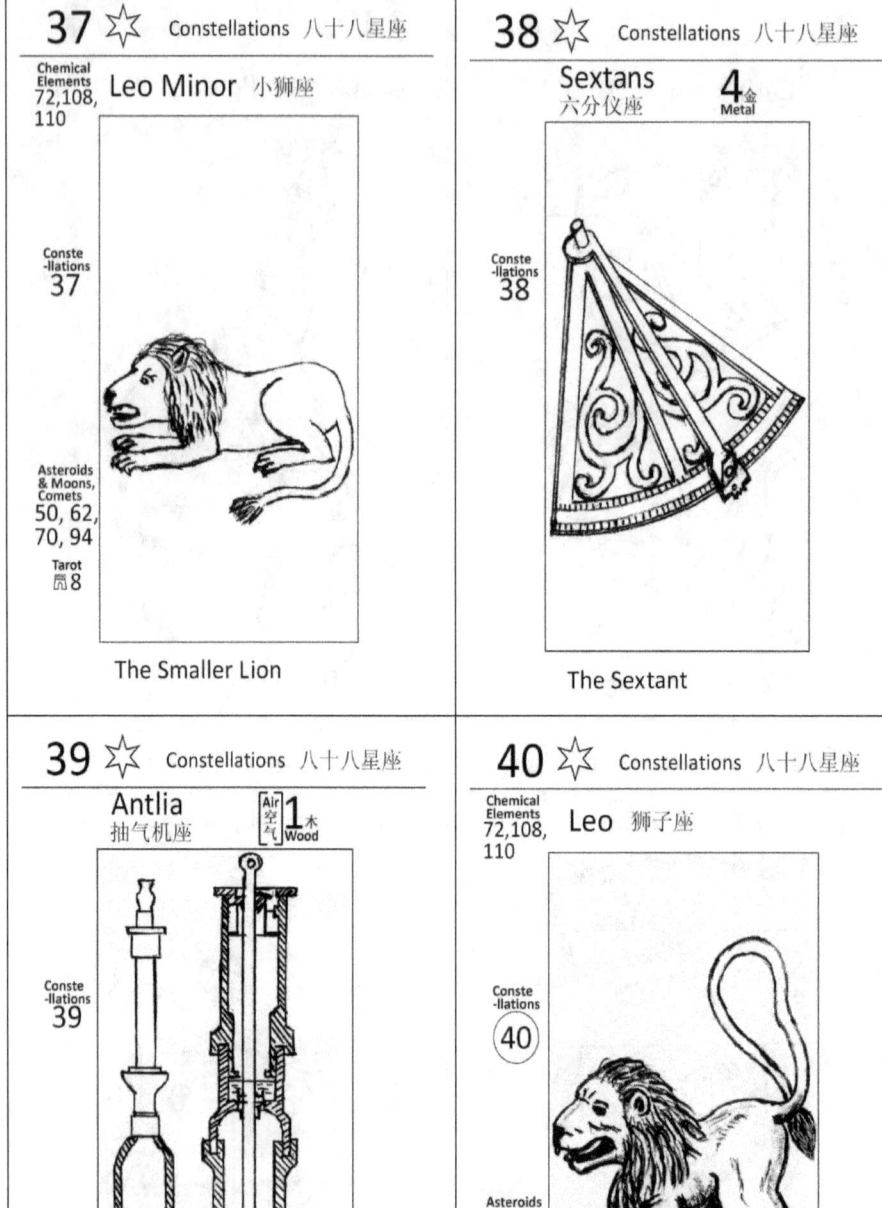

41 ☆ Constellations 八十八星座

Chamaeleon 蝘蜓座

Conste
-llations
41

The Chameleon

42 ☆ Constellations 八十八星座

Ursa Major 大熊座

Conste
-llations
42

Planets
& Moons
6/8,
6/24,
7/25

Asteroids
& Moons,
Comets
16

Tarot
♣9

The Larger Bear

43 ☆ Constellations 八十八星座

Crater 巨爵座 5水 Water

Conste
-llations
43

Lunar
Mansion
27

The Cup

44 ☆ Constellations 八十八星座

Hydra 长蛇座 5水 Water

Conste
-llations
44

Planets
& Moons
12/3

Lunar
Mansion
24, 25,
26, 27

The Water Monster

49 ☆ Constellations 八十八星座

Centaurus 半人马座

Conste
-llations
49

Asteroids
& Moons,
Comets
117,
120,
121

The Centaur Chiron

The Centaur

50 ☆ Constellations 八十八星座

Canes Venatici 猎犬座

Conste
-llations
50

Planets
& Moons
12/4

Asteroids
& Moons,
Comets
4

Lunar
Mansion
16

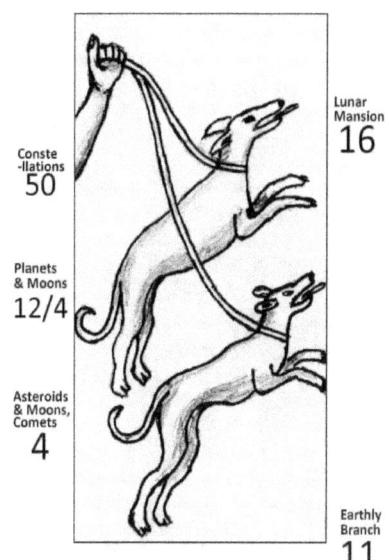

Earthly
Branch
11
戌狗
Dog

The Hunting Dogs

51 ☆ Constellations 八十八星座

Virgo 室女座

Conste
-llations
(51)

Asteroids
& Moons,
Comets
39

Zodiac
Sign
6
Virgo

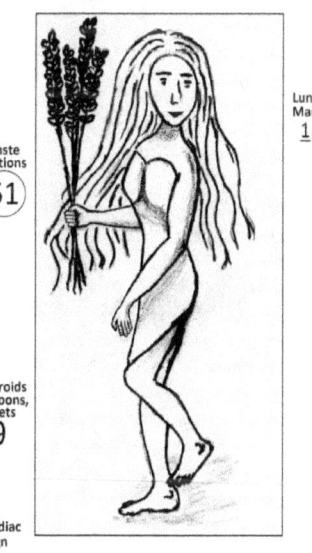

Lunar
Mansion
1, 2

The Virgin

52 ☆ Constellations 八十八星座

Circinus 圆规座

4 金
Metal

Conste
-llations
52

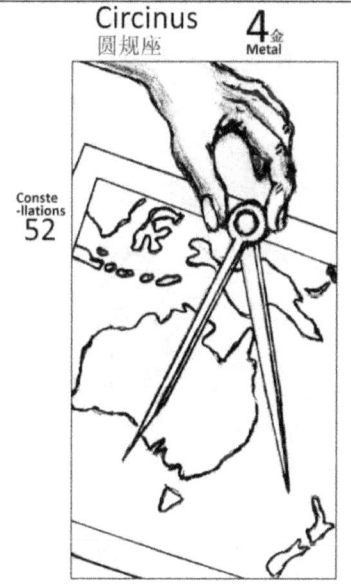

The Pair of Compasses

53 ✧ Constellations 八十八星座
Boötes 牧夫座

Constellations 53

The Herdsman

54 ✧ Constellations 八十八星座
Ursa Minor 小熊座

Constellations 54

Planets & Moons
6/8, 6/24, 7/25

Asteroids & Moons, Comets
16, 82

Tarot ♣9

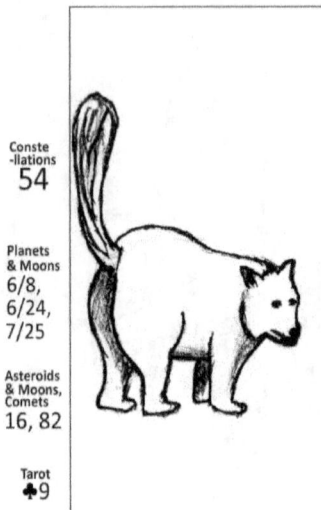

The Smaller Bear

55 ✧ Constellations 八十八星座
Draco 天龙座

Chemical Elements 48

Constellations 55

Lunar Mansions 1, 2

Tarot 筒22

Earthly Branch 5 辰龙 Dragon

The Dragon

56 ✧ Constellations 八十八星座
Libra 天秤座

Chemical Elements 78, 85

Constellations (56)

Lunar Mansion 3

Planets & Moons 21/1

Asteroids & Moons, Comets 84

Tarot ◊6

Zodiac Sign 7 Libra

The Balance

57 ☆ Constellations 八十八星座

Chemical Elements 74

Lupus 豺狼座

Lunar Mansion **15**

Conste-llations **57**

Planets & Moons
6/54,
7/29,
34, 39,
40, 46,
48, 57

Asteroids & Moons, Comets
97/1,
97/2

The Wolf

58 ☆ Constellations 八十八星座

Corona Borealis 北冕座

4 金 Metal

Conste-llations **58**

Asteroids & Moons, Comets
15, 28
68, 79

The Northern Crown

59 ☆ Constellations 八十八星座

Norma 矩尺座

Conste-llations **59**

The Carpenter's Square

60 ☆ Constellations 八十八星座

Triangulum Australe
南三角座

Conste-llations **60**

The Southern Triangle

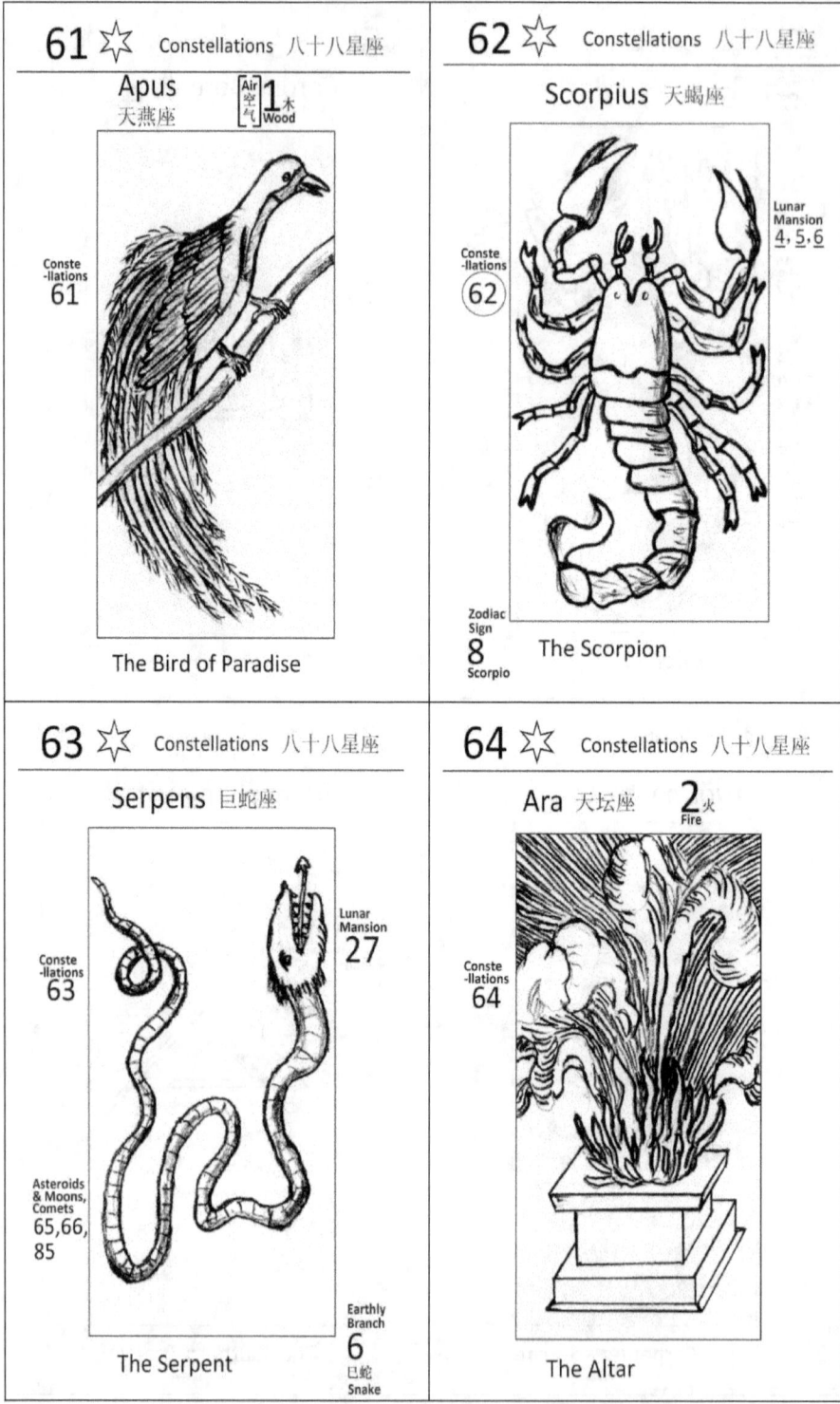

65 ✪ Constellations 八十八星座 ## Hercules 武仙座  Conste-llations **65** Asteroids & Moons, Comets **65** The Greek Hero	**66** ✪ Constellations 八十八星座 ## Ophiuchus 持蛇夫座 The Blameless Physician  Conste-llations **66** Asteroids & Moons, Comets **65, 66, 85** Lunar Mansion **27** Asclepius Earthly Branch **6** 巳蛇 Snake The Serpent-bearer
67 ✪ Constellations 八十八星座 ## Corona Austrina **4** 金 Metal 南冕座 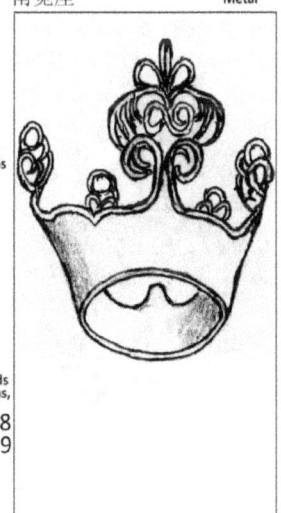 Conste-llations **67** Asteroids & Moons, Comets **15, 28, 68, 79** The Southern Crown	**68** ✪ Constellations 八十八星座 ## Scutum 盾牌座 The Shield of John III Sobieski of Poland 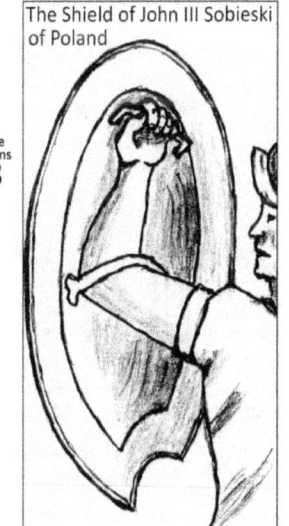 Conste-llations **68** The Shield

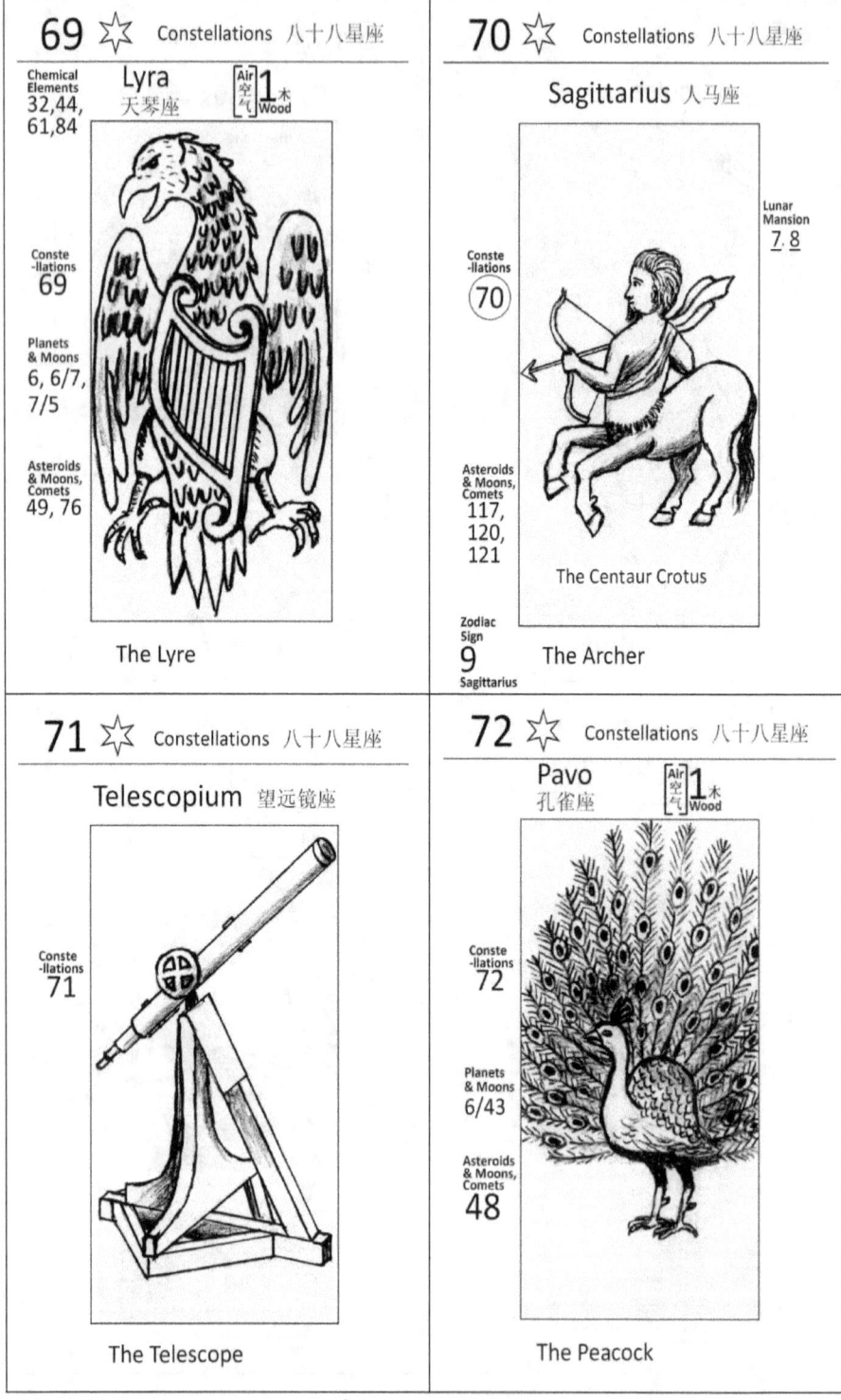

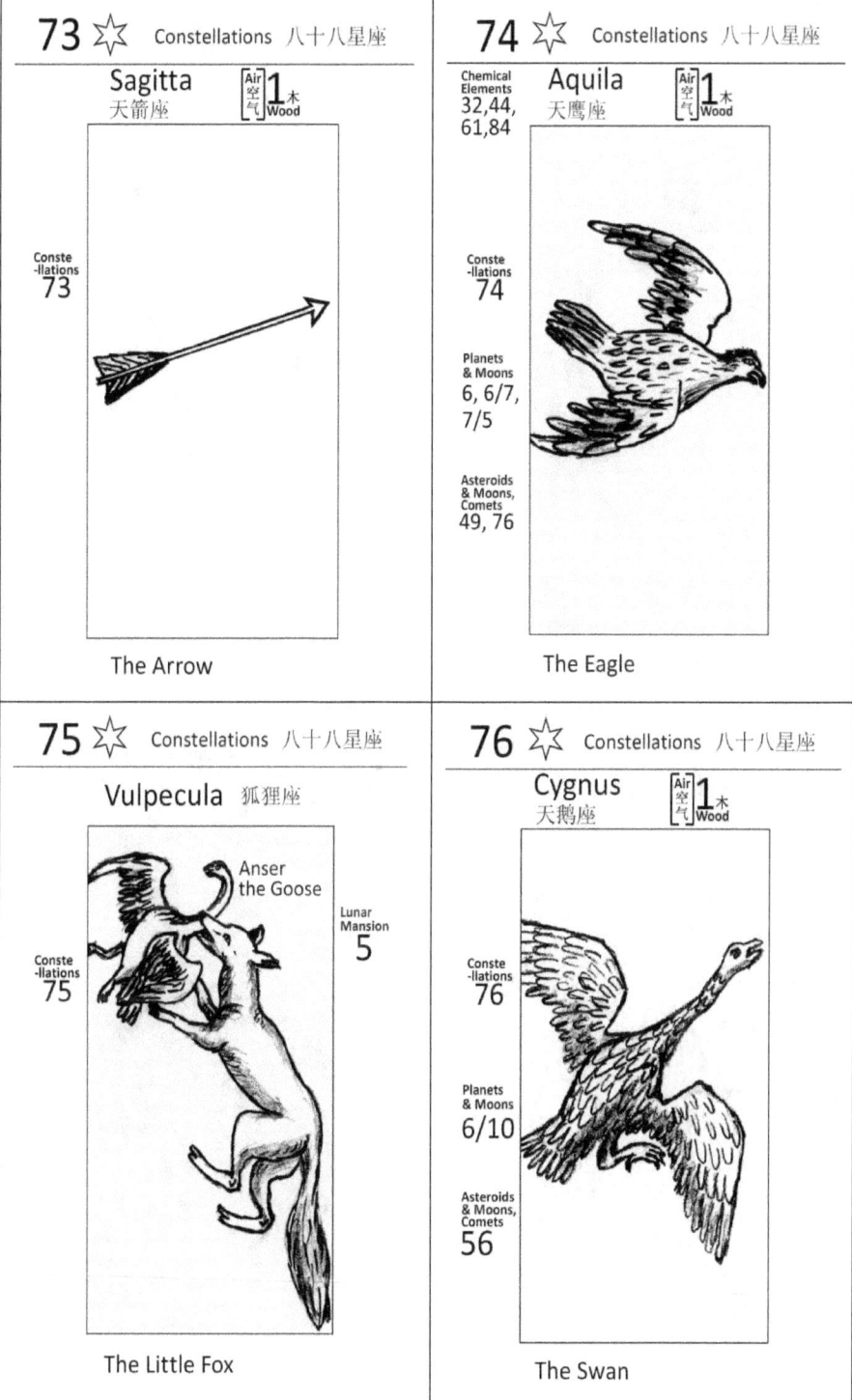

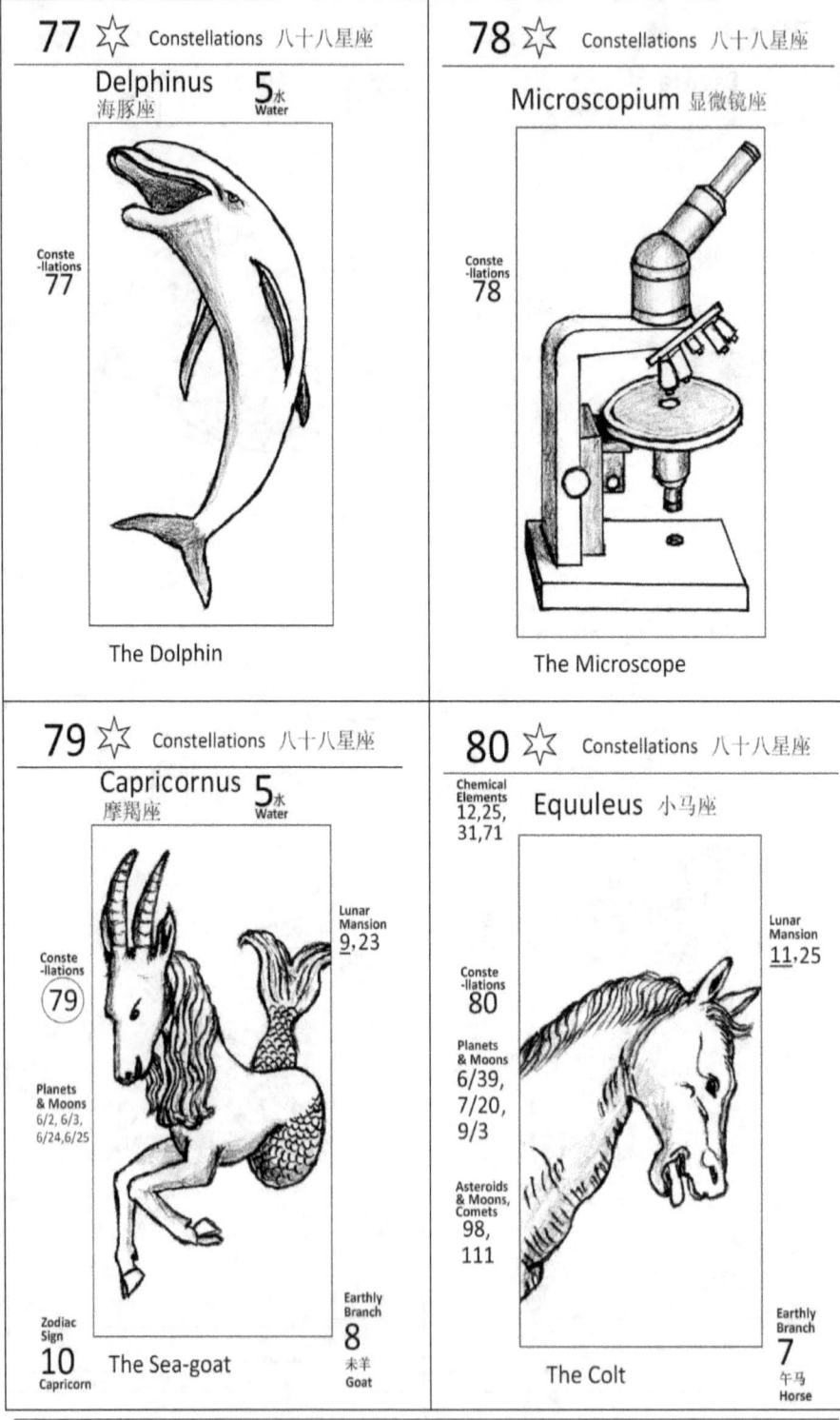

85 ☆ Constellations 八十八星座

Lacerta 蝎虎座

Conste-llations 85

The Lizard

86 ☆ Constellations 八十八星座

Pegasus 飞马座

Chemical Elements
12,25,
31,71

Conste-llations 86

Planets & Moons
6/39,
7/20,
9/3

Asteroids & Moons, Comets
98,
111

Lunar Mansion
$\underline{12},\underline{13},$
$\underline{14},25$

Earthly Branch
7
午马
Horse

The Winged Horse

87 ☆ Constellations 八十八星座

Octans
南极座

The Octant 八分仪

Conste-llations 87

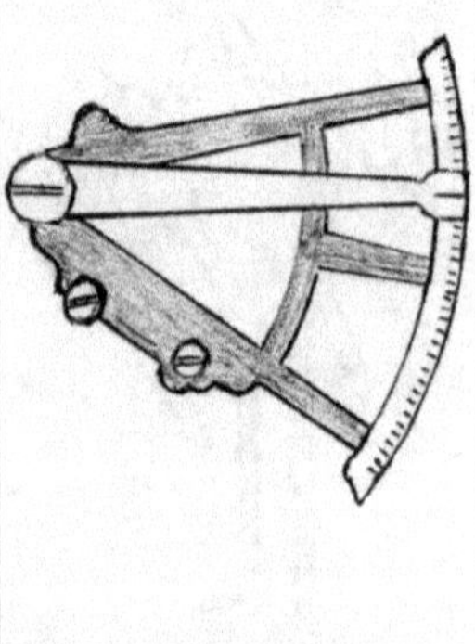

Constellation of the South Pole

88 ☆ Constellations 八十八星座

Tucana 杜鹃座

Air 空气 | **1** 木 Wood

Conste-llations 88

The Toucan

The 10 Suns of the Shang Dynasty 商朝十个太阳，十天为一周

The ten suns of the Shang Dynasty were represented by sun birds and are directly related to the 10 Heavenly Stems which were the names of the days of a 10 day week. The Fú Sāng tree (扶桑) which was thought to grow in the east, gave rest to these birds while each waited in turn to commence its flight across the sky carrying one of the ten suns of a ten day week. The western counterpart of the Fú Sāng 扶桑 tree was the Ruò Mù tree (若木) on which the sun birds rested after their long journey across the sky. The sun bird which survived the shooting by Hòu Yì is number 3 丙 Bǐng (阳火 Yáng Fire). This is the same sun as the Greco/Roman sun and the sun carried by Fú Xī (Chinese Luminaries 7). Some suns are related to suns, birds or animals among the 28 Lunar Mansions, 12 Earthly Branches, 88 Constellations as well as the Asteroids and Tarot cards.

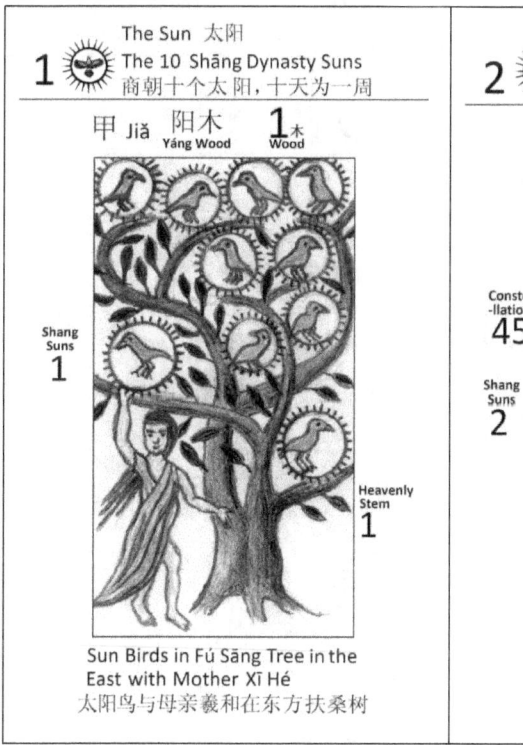

Sun Birds in Fú Sāng Tree in the East with Mother Xī Hé
太阳鸟与母亲羲和在东方扶桑树

太阳鸟和芙蓉花
Sun Bird and Red Hibiscus

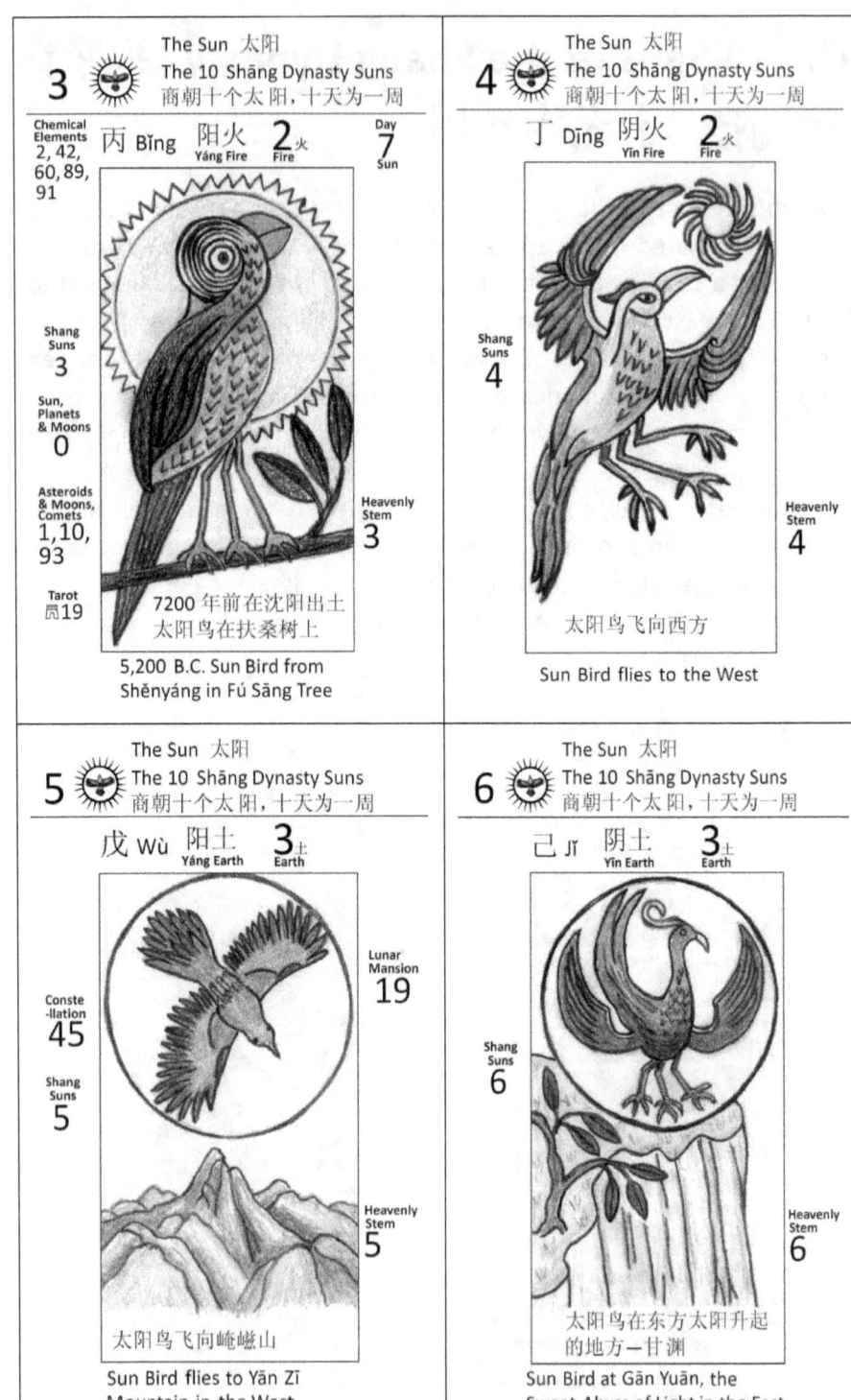

7 ☀ The Sun 太阳 The 10 Shāng Dynasty Suns 商朝十个太阳，十天为一周 庚 Gēng 阳金 Yáng Metal 4 金 Metal Constellation 45 Shang Suns 7 Lunar Mansion 19 Heavenly Stem 7  Sun Birds rest in Ruò Mù Tree in the West 太阳鸟在西方若木树上休息	**8** ☀ The Sun 太阳 The 10 Shāng Dynasty Suns 商朝十个太阳，十天为一周 辛 Xīn 阴金 Yīn Metal 4 金 Metal Shang Suns 8 Heavenly Stem 8  太阳鸟在西方若木树上 Sun Bird in Ruò Mù Tree in the West
9 ☀ The Sun 太阳 The 10 Shāng Dynasty Suns 商朝十个太阳，十天为一周 壬 Rén 阳水 Yáng Water 5 水 Water Shang Suns 9 Heavenly Stem 9  Sun Bird bathes in Gān Yuān, the Sweet Abyss of the East 太阳鸟在东方的甘渊沐浴	**10** ☀ The Sun 太阳 The 10 Shāng Dynasty Suns 商朝十个太阳，十天为一周 癸 Guǐ 阴水 Yīn Water 5 水 Water Shang Suns 10 Heavenly Stem 10 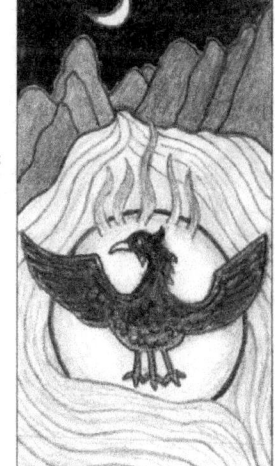 Sun Bird cools down in waters of Yù Gǔ Valley in the West 太阳鸟在西方的遇谷水中冷却

The Sun, Planets and Moons 太阳，行星和卫星

This section includes the Sun, all planets, dwarf planets and their moons as well as some potential dwarf planets and their moons between Mercury and Sedna. Relationships with days of the week, the Chinese 5 elements, the directions, the 88 Constellations, the 28 Lunar Mansions, the 12 Earthly Branches, Asteroids and their Moons, and Tarot cards are included.

Sun, Mercury, Venus, Earth, Mars, Ceres and their Moons 太阳，水星，金星，地球，火星，谷神星及其卫星

There are 10 cards in this section. Mercury, Venus and Ceres have no moons and Ceres is a dwarf planet. The sun is numbered 0 (zero).

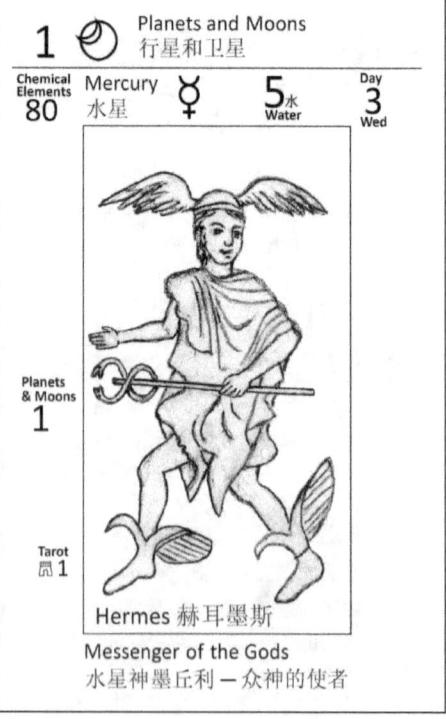

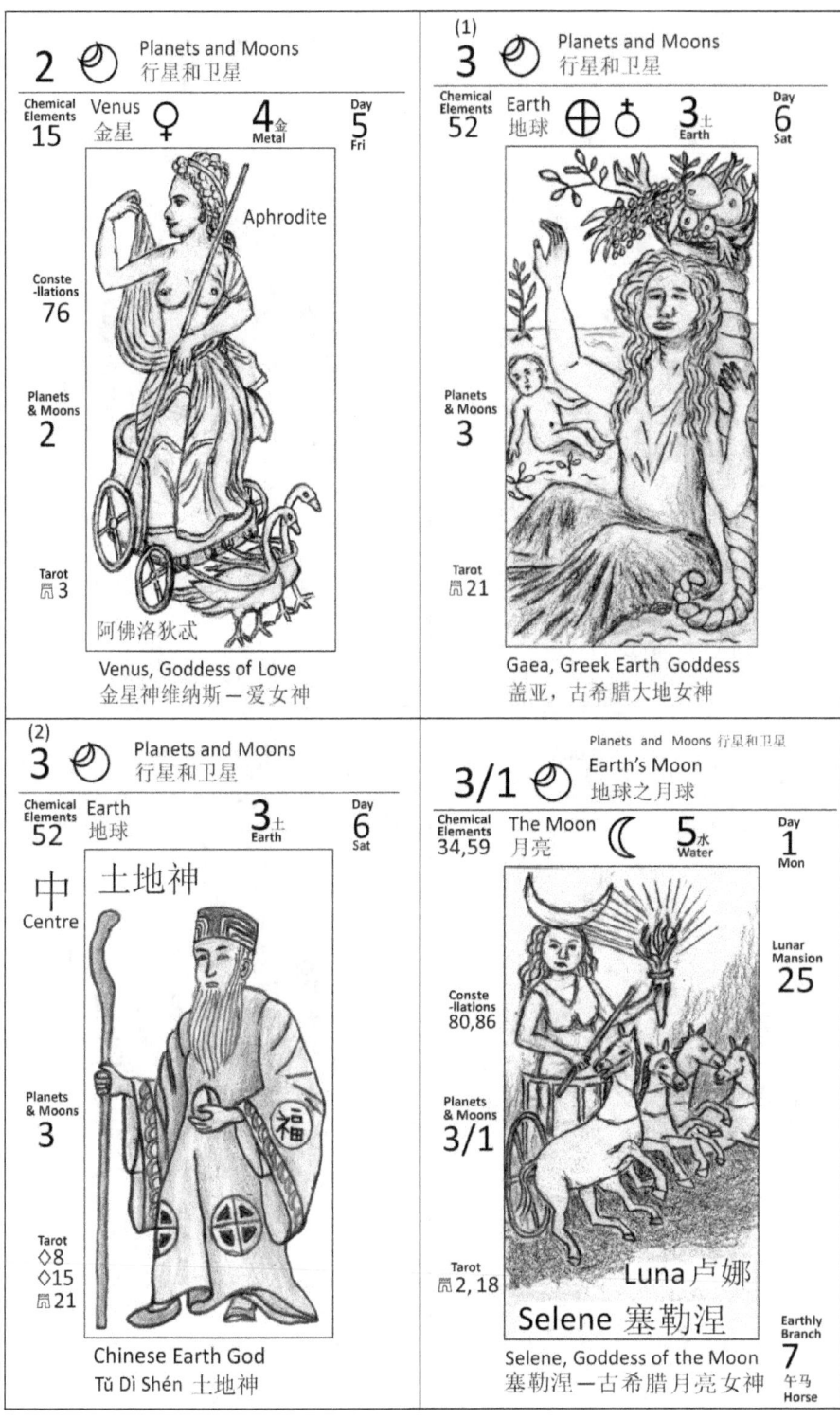

4 🌑 Planets and Moons
行星和卫星

Mars 火星 ♂ **2** 火 Fire **Day 2** Tue

Planets & Moons **4**

Tarot 16

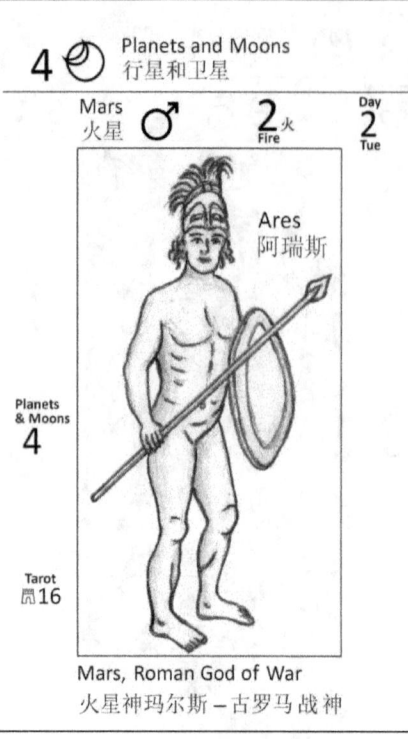

Ares 阿瑞斯

Mars, Roman God of War
火星神玛尔斯 — 古罗马战神

4/1 🌑 Planets and Moons 行星和卫星
The Moon of Mars 火星之卫星

Phobos 福波斯

Planets & Moons **4/1**

Fear   Timor

Attendant of Greek War God
Ares 古希腊战神阿瑞斯的副官

4/2 🌑 Planets and Moons 行星和卫星
The Moon of Mars 火星之卫星

Deimos 戴摩斯

Planets & Moons **4/2**

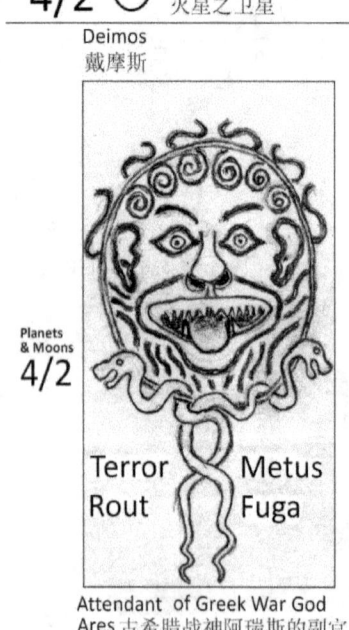

Terror   Metus
Rout   Fuga

Attendant of Greek War God
Ares 古希腊战神阿瑞斯的副官

---

5 🌑 Planets and Moons
行星和卫星

Chemical Elements **58**

Ceres 谷神星 ⚳

Demeter 得墨忒耳

Constellations 8, 63, 66

Planets & Moons **5**

Lunar Mansion **27**

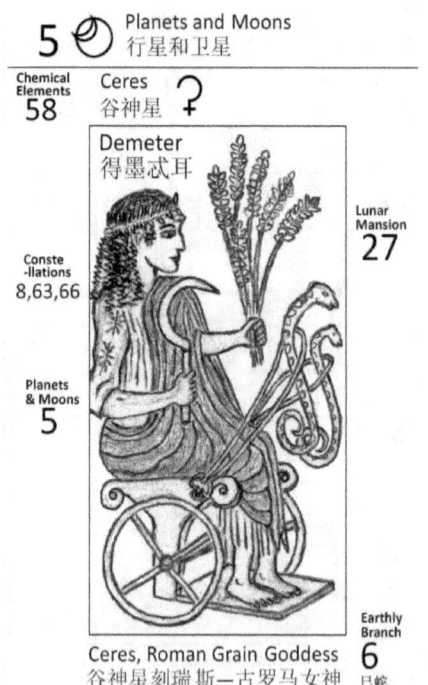

Earthly Branch **6** 巳蛇 Snake

Ceres, Roman Grain Goddess
谷神星刻瑞斯 — 古罗马女神

Planets and Moons 行星和卫星

This section includes all planets, dwarf planets and their moons as well as some potential dwarf planets and their moons between Mercury and Sedna. Relationships with days of the week, the Chinese 5 elements, the directions, the 88 Constellations, the 28 Lunar Mansions, the 12 Earthly Branches, Asteroids and their Moons, and Tarot cards are included.

Jupiter and its Moons
木星及其卫星

Jupiter has 67 moons but some have not yet been named by astronomers and are represented by a blank card.

6/6 🌙 Planets and Moons 行星和卫星
The Moons of Jupiter 木星之卫星

Europa 欧罗巴 5 水 Water

Constellation 16
Planets & Moons 6/6
Asteroids & Moons, Comets 80
Tarot 27

Lunar Mansion 9
Earthly Branch 2 丑牛 Ox

Phoenician Princess carried by Bull to Crete

6/7 🌙 Planets and Moons 行星和卫星
The Moons of Jupiter 木星之卫星

Chemical Elements 32,44, 61,84
Ganymede 伽倪墨得斯 Air 空气 1 木 Wood

Constellation 69,74
Planets & Moons 6/7
Asteroids & Moons, Comets 49, 76

Cup-bearer of Zeus abducted by an Eagle

6/8 🌙 Planets and Moons 行星和卫星
The Moons of Jupiter 木星之卫星

Chemical Elements 59,98
Callisto 卡利斯托

Constellation 42, 54
Planets & Moons 6/8
Asteroids & Moons, Comets 16, 82

Lover of Zeus becomes a She-bear

6/9 🌙 Planets and Moons 行星和卫星
The Moons of Jupiter 木星之卫星

Themisto 忒弥斯托 5 水 Water

Planets & Moons 6/9

Lover of Zeus; Mother of the River Danube

6/13 🌑 Planets and Moons 行星和卫星 The Moons of Jupiter 木星之卫星 Elara 伊拉拉 **3** 土 Earth  Lover of Zeus; Mother of the Giant Tityus	**6/14** 🌑 Planets and Moons 行星和卫星 The Moons of Jupiter 木星之卫星 Dia 黛  Wife of Ixion; Lover of Zeus; Mother of Pirithous
6/15 🌑 Planets and Moons 行星和卫星 The Moons of Jupiter 木星之卫星 Carpo 卡尔波 Air 空气 **1** 木 Wood Thallo, Auxo — Spring, Summer Carpo — Fruit, Autumn First Generation of Horae (the Hours, the Seasons)	**6/16** 🌑 Planets and Moons 行星和卫星 The Moons of Jupiter 木星之卫星 S/2003 J 12

6/17 ☾	Planets and Moons 行星和卫星 The Moons of Jupiter 木星之卫星	6/18 ☾	Planets and Moons 行星和卫星 The Moons of Jupiter 木星之卫星
Euporie 欧波里亚 Air 空气 1 木 Wood  Euporie, Orthosie, Pherusa Abundance, Prosperity, Farms Third Generation of Horae (the Hours, the Seasons)		S/2003 J 3	
6/19 ☾	Planets and Moons 行星和卫星 The Moons of Jupiter 木星之卫星	**6/20** ☾	Planets and Moons 行星和卫星 The Moons of Jupiter 木星之卫星
S/2003 J 18		S/2011 J 1	

6/21
Planets and Moons 行星和卫星
The Moons of Jupiter
木星之卫星

S/2010 J 2

6/22
Planets and Moons 行星和卫星
The Moons of Jupiter
木星之卫星

Thelxinoe
忒尔克西诺厄

Delighting the Heart

Planets & Moons
6/22

Tarot
♣18

Thelxinoe, Aoede, Arche, Melete

One of Four Muses, Daughters of Zeus

6/23
Planets and Moons 行星和卫星
The Moons of Jupiter
木星之卫星

Euanthe
欧安忒

Planets & Moons
6/23

Mother of the Charites or Graces by Zeus

6/24
Planets and Moons 行星和卫星
The Moons of Jupiter
木星之卫星

Chemical Elements
98

Helike
赫利刻

3 土
Earth

Conste-llation
42, 54

Planets & Moons
6/24

Asteroids & Moons, Comets
82

Lunar Mansion
23

Zodiac Sign
10
Capricorn

Helike (Helice) or Adrasteia changed into a She-bear

Earthly Branch
8
未羊
Goat

6/25 — The Moons of Jupiter 木星之卫星
Planets and Moons 行星和卫星

Orthosie 俄尔托西亚
Air 空气 / 1 木 Wood
Orthosie Prosperity

Euporie, Pherusa

Constellation 24, 79

Planets & Moons 6/25

Lunar Mansion 23

Earthly Branch 8 未羊 Goat

Zodiac Sign 10 Capricorn

Third Generation of Horae (the Hours, the Seasons)

6/26 — The Moons of Jupiter 木星之卫星
Planets and Moons 行星和卫星

Iocaste 伊俄卡斯忒

Planets & Moons 6/26

Jocasta 约卡斯塔
Mother of Oedipus

6/27 — The Moons of Jupiter 木星之卫星
Planets and Moons 行星和卫星

S/2003 J 16

6/28 — The Moons of Jupiter 木星之卫星
Planets and Moons 行星和卫星

Praxidike 普剌克西狄刻

Planets & Moons 6/28

Goddess of Punishment;
Exactor of Justice

Astronomical Pencil Drawings - John Oxenham Goodman

6/29 — The Moons of Jupiter 木星之卫星

Harpalyke 哈耳帕吕刻

Air 空气 / 1 木 Wood

Planets & Moons 6/29

Incestuous Murderer turned into a Bird of Prey

6/30 — The Moons of Jupiter 木星之卫星

Mneme 谟涅墨

Memory

Planets & Moons 6/30

Tarot ♣18

One of the 3 original Muses: Mneme, Aoede, Melete

6/31 — The Moons of Jupiter 木星之卫星

Hermippe 赫耳弥珀

Planets & Moons 6/31

Mother of King Orchomenus and Consort of Zeus

6/32 — The Moons of Jupiter 木星之卫星

Thyone 堤俄涅

Lunar Mansion 16

Conste-llation 25, 29, 50

Planets & Moons 6/32

Asteroids & Moons, Comets 4

Dionysus carries his mother to Mount Olympus

Semele becomes the Goddess Thyone

Earthly Branch 11 戌狗 Dog

6/33 🌑 Planets and Moons 行星和卫星 **The Moons of Jupiter** 木星之卫星 Ananke 阿南刻 Planets & Moons **6/33**  Anance Necessity, Destiny; Mother of the Fates	**6/34** 🌑 Planets and Moons 行星和卫星 **The Moons of Jupiter** 木星之卫星 Herse 赫斯 Conste-llation 8, 63, 66 Planets & Moons **6/34** Asteroids & Moons, Comets 65, 66, 85  Lunar Mansion **27** Earthly Branch **6** 巳蛇 Snake Herse and Aglauros, daughters of Zeus and Selene
6/35 🌑 Planets and Moons 行星和卫星 **The Moons of Jupiter** 木星之卫星 Aitne 埃特那 **2** 火 Fire Planets & Moons **6/35**  The Giant Enceladus buried under Aitne's Mountain	**6/36** 🌑 Planets and Moons 行星和卫星 **The Moons of Jupiter** 木星之卫星 Kale 卡勒伊斯 Planets & Moons **6/36**  Cale Goddess of Beauty; One of the Graces

6/37 Planets and Moons 行星和卫星
The Moons of Jupiter
木星之卫星

Chemical Elements **59**

Taygete
卡勒伊斯

Lunar Mansion **26**

Planets & Moons
6/37

Tarot
♠21, 22

One of the 7 Pleiades;
Daughters of Atlas

6/38 Planets and Moons 行星和卫星
The Moons of Jupiter
木星之卫星

S/2003 J 19

6/39 Planets and Moons 行星和卫星
The Moons of Jupiter
木星之卫星

Chaldene
卡尔得涅

Conste-llation
80, 86

Lunar Mansion **25**

Planets & Moons
6/39

Asteroids & Moons, Comets
98, 111

Tarot
♠12
♣12
♦12
♥12

Chaldene
Solymos

Mother of Solymos, founder
of the city of Termessos

Earthly Branch **7**
午马
Horse

6/40 Planets and Moons 行星和卫星
The Moons of Jupiter
木星之卫星

S/2003 J 15

6/41 Planets and Moons 行星和卫星 The Moons of Jupiter 木星之卫星 S/2003 J10	**6/42** Planets and Moons 行星和卫星 The Moons of Jupiter 木星之卫星 S/2003 J 23

6/43 Planets and Moons 行星和卫星
The Moons of Jupiter 木星之卫星
Erinome 厄里诺墨   Air 空气 1 木 Wood

Constellation 72
Planets & Moons 6/43
Asteroids & Moons, Comets 12

Changed into a peahen after intimacy with Adonis

6/44 Planets and Moons 行星和卫星
The Moons of Jupiter 木星之卫星
Aoede 阿俄伊得

Planets & Moons 6/44
Tarot ♣18

"Song", one of the 3 original Muses

6/45 🌓 Planets and Moons 行星和卫星 The Moons of Jupiter 木星之卫星 Kallichore 卡利科瑞　　3 土 Earth  Planets & Moons 6/45 Nurse of Dionysus on Mount Nysa	**6/46** 🌓 Planets and Moons 行星和卫星 The Moons of Jupiter 木星之卫星 Kalyke 卡吕刻　　 1 木 Wood  Planets & Moons 6/46 Asteroids & Moons, Comets 11 Silenus holds Dionysus; Kalike gives him grapes
6/47 🌓 Planets and Moons 行星和卫星 The Moons of Jupiter 木星之卫星 Carme 加尔尼　　3 土 Earth  Planets & Moons 6/47 Cretan Harvest Nymph	**6/48** 🌓 Planets and Moons 行星和卫星 The Moons of Jupiter 木星之卫星 Callirrhoe 卡利罗厄　　5 水 Water  Planets & Moons 6/48 Daughter of the River God Achelous

6/49 ☽ Planets and Moons 行星和卫星
The Moons of Jupiter
木星之卫星

Eurydome
欧律多墨

Planets & Moons
6/49

Mother of the Charites or Graces; Rival claimant

6/50 ☽ Planets and Moons 行星和卫星
The Moons of Jupiter
木星之卫星

S/2011 J 2

6/51 ☽ Planets and Moons 行星和卫星
The Moons of Jupiter
木星之卫星

Pasithee
帕西忒亚

Planets & Moons
6/51

Goddess of Relaxation with husband Hypnos "Sleep"

6/52 ☽ Planets and Moons 行星和卫星
The Moons of Jupiter
木星之卫星

S/2010 J 1

6/53 ☽ Planets and Moons 行星和卫星
The Moons of Jupiter 木星之卫星

Kore
科瑞

Conste-llation
25,29, 50

Planets & Moons
6/53

Asteroids & Moons, Comets
4

Lunar Mansion
16

Earthly Branch
11
戌狗
Dog

Saved from abduction when Cerberus killed Pirithous

6/54 ☽ Planets and Moons 行星和卫星
The Moons of Jupiter 木星之卫星

Chemical Elements
74

Cyllene
库勒涅

Conste-llation
57

Planets & Moons
6/54

Asteroids & Moons, Comets
97/1, 97/2

Lunar Mansion
15

Nymph whose son Lycaon was turned into a wolf

6/55 ☽ Planets and Moons 行星和卫星
The Moons of Jupiter 木星之卫星

Eukelade
欧刻拉得

Planets & Moons
6/55

Tarot
♣18

One of the lesser-known Muses, daughters of Zeus

6/56 ☽ Planets and Moons 行星和卫星
The Moons of Jupiter 木星之卫星

S/2003 J4

6/57 Planets and Moons 行星和卫星
The Moons of Jupiter 木星之卫星

Pasiphaë
帕西菲

Constellation 16

Lunar Mansion 9

Planets & Moons 6/57

Earthly Branch 2 丑牛 Ox

Mother of the Minotaur

6/58 Planets and Moons 行星和卫星
The Moons of Jupiter 木星之卫星

Hegemone
赫革摩涅

Planets & Moons 6/58

Queen or Leader of the Charites or Graces

6/59 Planets and Moons 行星和卫星
The Moons of Jupiter 木星之卫星

Arche
阿耳刻

Planets & Moons 6/59

The 4th Muse, the "Beginning"

6/60 Planets and Moons 行星和卫星
The Moons of Jupiter 木星之卫星

Isonoe
伊索诺厄

5 水 Water

Planets & Moons 6/60

Lover of Zeus who became a spring of water

6/61	Planets and Moons 行星和卫星 The Moons of Jupiter 木星之卫星

S/2003 J 9

6/62	Planets and Moons 行星和卫星 The Moons of Jupiter 木星之卫星

S/2003 J 5

6/63	Planets and Moons 行星和卫星 The Moons of Jupiter 木星之卫星

Sinope
希诺佩

5 水 Water

Planets & Moons
6/63

Taken by ship to Anatolia;
Founds city on Black Sea

6/64	Planets and Moons 行星和卫星 The Moons of Jupiter 木星之卫星

Sponde
斯蓬德

5 水 Water

Planets & Moons
6/64

Pouring a drink offering on the altar

6/65 ☾

Planets and Moons 行星和卫星
The Moons of Jupiter
木星之卫星

Autonoe
奥托诺厄

Lunar Mansion
7

Planets & Moons
6/65

Joined Dionysus in revelry and orgiastic rites

6/66 ☾

Planets and Moons 行星和卫星
The Moons of Jupiter
木星之卫星

Chemical Elements
2, 42, 60, 89, 91

Megaclite
墨伽克利忒

Sun
3/2

Planets & Moons
6/66

Asteroids & Moons, Comets
1, 10, 93

Megaclite with grandparents Sun God Helios and Sea Nymph Rhodos

6/67 ☾

Planets and Moons 行星和卫星
The Moons of Jupiter
木星之卫星

S/2003 J 2

Planets and Moons 行星和卫星

This section includes all planets, dwarf planets and their moons as well as some potential dwarf planets and their moons between Mercury and Sedna. Relationships with days of the week, the Chinese 5 elements, the directions, the 88 Constellations, the 28 Lunar Mansions, the 12 Earthly Branches, Asteroids and Moons and Tarot cards are included.

Saturn and its Moons
土星及其卫星

Saturn has 62 moons but some have not yet been named by astronomers and are represented by a blank card.

7/2 ☾ Planets and Moons 行星和卫星 The Moons of Saturn 土星之卫星 Pan 潘 Planets & Moons **7/2** Tarot ♣18  Lustful Fertility God of mountains and forests	**7/3** ☾ Planets and Moons 行星和卫星 The Moons of Saturn 土星之卫星 Daphnis 达菲尼斯 Planets & Moons **7/3** Tarot ♣18 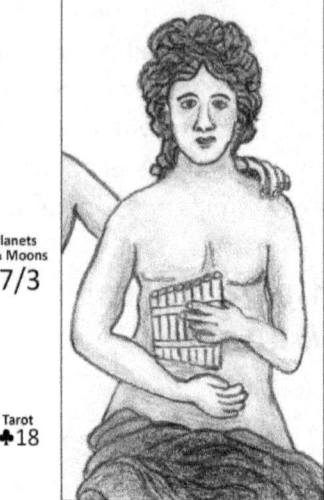 Blind Bisexual Sicilian Musician
7/4 ☾ Planets and Moons 行星和卫星 The Moons of Saturn 土星之卫星 Atlas 阿特拉斯 Planets & Moons **7/4**  Supporter of the Heavens	**7/5** ☾ Planets and Moons 行星和卫星 The Moons of Saturn 土星之卫星 Prometheus 普罗米修斯 Chemical Elements 32,44, 61,84 Conste-llation 69,74 Planets & Moons **7/5**  Benefactor of Mankind punished by Zeus

7/6	Planets and Moons 行星和卫星 **The Moons of Saturn** 土星之卫星

Pandora
潘多拉

Planets & Moons
7/6

Asteroids & Moons, Comets
60

First Mortal Woman

7/7	Planets and Moons 行星和卫星 **The Moons of Saturn** 土星之卫星

Epimetheus
厄庇墨透斯

Planets & Moons
7/7

Husband of Pandora

7/8	Planets and Moons 行星和卫星 **The Moons of Saturn** 土星之卫星

Janus
雅努斯

Planets & Moons
7/8

Roman God of Doors, Gates, Beginnings and Endings

7/9	Planets and Moons 行星和卫星 **The Moons of Saturn** 土星之卫星

Aegaeon
埃该翁

 5 水 Water

Planets & Moons
7/9

Giant of Aegean Sea with 100 arms also called Briareus
布里阿瑞俄斯

7/10 ◐ Planets and Moons 行星和卫星
The Moons of Saturn 土星之卫星

Mimas 埃该翁

Giant killed by the gods and buried under a mountain

7/11 ◐ Planets and Moons 行星和卫星
The Moons of Saturn 土星之卫星

Methone 墨托涅 — 5 水 Water

Changed into a kingfisher by Sea Goddess Amphitrite

7/12 ◐ Planets and Moons 行星和卫星
The Moons of Saturn 土星之卫星

Anthe 安忒 — 5 水 Water

Changed into a kingfisher when she jumped into the sea

7/13 ◐ Planets and Moons 行星和卫星
The Moons of Saturn 土星之卫星

Pallene 帕勒涅 — 5 水 Water

Changed into a kingfisher as she jumped into the sea

7/14 Planets and Moons 行星和卫星
The Moons of Saturn 土星之卫星
Enceladus
恩克拉多斯

Planets & Moons
7/14

Giant defeated by Athena

7/15 Planets and Moons 行星和卫星
The Moons of Saturn 土星之卫星
Tethys
忒堤斯

5 Water

Planets & Moons
7/15

Wife of Oceanus, Mother of ocean nymphs and rivers

7/16 Planets and Moons 行星和卫星
The Moons of Saturn 土星之卫星
Telesto
忒勒斯托

5 Water

Planets & Moons
7/16

Ocean Nymph personifying Success

7/17 Planets and Moons 行星和卫星
The Moons of Saturn 土星之卫星
Calypso
卡吕普索

5 Water

Planets & Moons
7/17

Asteroids & Moons, Comets
44

Calypso finds Odysseus on the seashore

7/18 Planets and Moons 行星和卫星
The Moons of Saturn 土星之卫星

Dione 狄俄涅

Air 空气 | 1 木 Wood

Conste-llation **23**

Planets & Moons **7/18**

Asteroids & Moons, Comets **90**

Presiding Goddess at Oracular Shrine of Dodona

7/19 Planets and Moons 行星和卫星
The Moons of Saturn 土星之卫星

Helene 海伦

Planets & Moons **7/19**

Wife of Menelaus abducted by Paris Prince of Troy

7/20 Planets and Moons 行星和卫星
The Moons of Saturn 土星之卫星

Polydeuces 波吕丢刻斯

Pollux 波鲁克斯

Lunar Mansion **25**

Conste-llation **80, 86**

Planets & Moons **7/20**

Asteroids & Moons, Comets **98, 111**

Tarot
♠12
♣12
♦12
♥12

Earthly Branch **7** 午马 Horse

Immortal son of Zeus and twin brother of mortal Castor

7/21 Planets and Moons 行星和卫星
The Moons of Saturn 土星之卫星

Rhea 瑞亚

Planets & Moons **7/21**

Asteroids & Moons, Comets **94**

Rhea hands Cronos a stone wrapped in baby clothes

7/22 — The Moons of Saturn 土星之卫星
Planets and Moons 行星和卫星

Titan 泰坦星

Chemical Elements: 22

Planets & Moons: 7/22

Son of Uranus "Heaven" and Gaea "Earth"

7/23 — The Moons of Saturn 土星之卫星
Planets and Moons 行星和卫星

Hyperion 许珀里翁

Chemical Elements: 2, 42, 60, 89, 91

2 火 Fire

Shang Suns: 3

Planets & Moons: 7/23

Father of Sun, Moon and Dawn helps support Uranus "Sky"

7/24 — The Moons of Saturn 土星之卫星
Planets and Moons 行星和卫星

Iapetus 伊阿珀托斯

Lunar Mansion: 16

Constellation: 25, 29, 50

Planets & Moons: 7/24

Asteroids & Moons, Comets: 4

Earthly Branch: 11 戌狗 Dog

Father of Atlas holds up the western sky

7/25 — The Moons of Saturn 土星之卫星
Planets and Moons 行星和卫星

Kiviuq 基维尤克

Chemical Elements: 98

5 水 Water

Planets & Moons: 7/25

Asteroids & Moons, Comets: 82

Wandering Inuit hero with shamanic powers

7/26 Planets and Moons 行星和卫星
The Moons of Saturn 土星之卫星
Ijiraq 伊耶拉克

Planets & Moons 7/26

Inuit half-human spirit seen on hazy days

7/27 Planets and Moons 行星和卫星
The Moons of Saturn 土星之卫星
Phoebe 福柏

Air 空气 1 木 Wood

Planets & Moons 7/27

Third Presiding Goddess at Oracle of Delphi

7/28 Planets and Moons 行星和卫星
The Moons of Saturn 土星之卫星
Paaliaq 波里阿科

Planets & Moons 7/28

Masked Inuit Shaman

7/29 Planets and Moons 行星和卫星
The Moons of Saturn 土星之卫星

Chemical Elements 74

Skathi 斯卡蒂

5 水 Water

Constellation 57

Lunar Mansion 15

Planets & Moons 7/29

Asteroids & Moons, Comets 97/1, 97/2

Norse Giantess and Goddess of Winter

7/30 🌒 Planets and Moons 行星和卫星 The Moons of Saturn 土星之卫星 Albiorix 阿尔比俄里克斯 **2** 火 Fire Conste-llation 25, 29, 50 Lunar Mansion **16** Planets & Moons 7/30 Earthly Branch **11** 戌狗 Dog Albiorix (Teutates), God of the Albici, carries victim to altar	**7/31** 🌒 Planets and Moons 行星和卫星 The Moons of Saturn 土星之卫星 S/2007 S 2
7/32 🌒 Planets and Moons 行星和卫星 The Moons of Saturn 土星之卫星 Bebhionn 贝芬 **5** 水 Water Planets & Moons 7/32 Giantess killed on the shores of Ireland by the giant Aeda	**7/33** 🌒 Planets and Moons 行星和卫星 The Moons of Saturn 土星之卫星 Erriapus 厄里阿波 Air 空气 **1** 木 Wood Planets & Moons 7/33 God of green vegetation and foliage

7/34 Planets and Moons 行星和卫星
The Moons of Saturn 土星之卫星

Sköll 斯库尔

2 火 Fire

Chemical Elements 2, 42, 60, 74, 89, 91

Lunar Mansion 15

Conste-llation 57

Sun 3/2

Planets & Moons 7/34

Asteroids & Moons, Comets 97/1, 97/2

Norse wolf that chases the sun

7/35 Planets and Moons 行星和卫星
The Moons of Saturn 土星之卫星

Siarnaq 西阿尔那克

5 水 Water

Planets & Moons 7/35

Sedna 赛德娜

Inuit Sea Goddess and mother of all sea creatures

7/36 Planets and Moons 行星和卫星
The Moons of Saturn 土星之卫星

Tarqeq 塔尔科克

5 水 Water

Chemical Elements 2, 34

Shang Suns 3

Planets & Moons 7/36

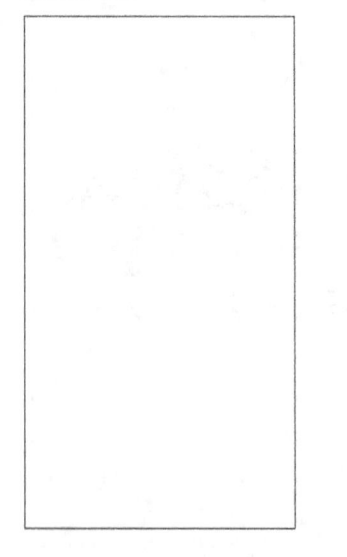

Inuit Moon God Tarqeq chases Sun Goddess Malina

7/37 Planets and Moons 行星和卫星
The Moons of Saturn 土星之卫星

S/2004 S 13

7/38 — The Moons of Saturn 土星之卫星
Planets and Moons 行星和卫星

Greip 格蕾普

3 Earth 土

Planets & Moons 7/38

Greip and Gjalp about to be crushed by Thor's chair

7/39 — The Moons of Saturn 土星之卫星
Planets and Moons 行星和卫星

Chemical Elements 74

Hyrrokkin 希尔罗金

Constellation 57

Lunar Mansion 15, 27

Planets & Moons 7/39

Asteroids & Moons, Comets 97/1, 97/2

Hyrrokkin launches Balder's funeral ship

7/40 — The Moons of Saturn 土星之卫星
Planets and Moons 行星和卫星

Chemical Elements 74

Jarnsaxa 雅恩莎撒

Constellation 57

Lunar Mansion 15

Planets & Moons 7/40

Asteroids & Moons, Comets 97/1, 97/2

Giantess; Lover of Thor

7/41 — The Moons of Saturn 土星之卫星
Planets and Moons 行星和卫星

Tarvos 塔沃斯

Air 空气 1 Wood 木

Constellation 16, 84

Lunar Mansion 9

Planets & Moons 7/41

Tarot ♠27 ♢21, 22 ♡2

Tarvos supports 3 cranes while Esus prunes a tree

Earthly Branch 2 丑牛 Ox

7/42 — The Moons of Saturn 土星之卫星
Planets and Moons 行星和卫星

Mundilfäri 蒙迪尔法利

Chemical Elements: 2, 34
Shang Suns: 3
Planets & Moons: 7/42

Frost Giant who turns cosmic mill; Father of Sun and Moon

7/43 — The Moons of Saturn 土星之卫星
Planets and Moons 行星和卫星

S/2006 S 1

7/44 — The Moons of Saturn 土星之卫星
Planets and Moons 行星和卫星

S/2004 S 17

7/45 — The Moons of Saturn 土星之卫星
Planets and Moons 行星和卫星

Bergelmir 贝格尔米尔

5 Water 水

Planets & Moons: 7/45

Bergelmir hides from flood with wife in hollow tree trunk

7/46 Planets and Moons 行星和卫星
The Moons of Saturn 土星之卫星

Chemical Elements 74

Narvi 那维

Constellation 57

Lunar Mansion 15

Planets & Moons 7/46

Asteroids & Moons, Comets 97/1, 97/2

Killed by his brother Vali who was changed into a wolf

7/47 Planets and Moons 行星和卫星
The Moons of Saturn 土星之卫星

Suttungr 苏图恩

5 水 Water

Planets & Moons 7/47

Suttungr bargains with the dwarfs for the magic mead

7/48 Planets and Moons 行星和卫星
The Moons of Saturn 土星之卫星

Chemical Elements 34, 59, 74

Hati 哈提

5 水 Water

Planets & Moons 7/48

Asteroids & Moons, Comets 97/1, 97/2

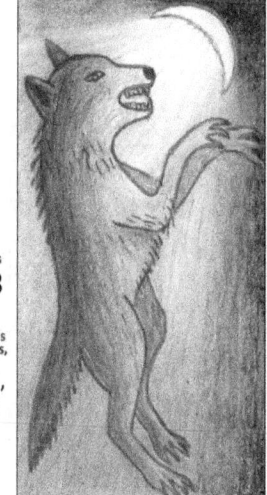

Norse wolf that chases the moon

7/49 Planets and Moons 行星和卫星
The Moons of Saturn 土星之卫星

S/2004 S 12

7/50 Planets and Moons 行星和卫星
The Moons of Saturn 土星之卫星

Farbauti 法布提

Chemical Elements: 90

2 Fire 火

Day 4 Thu

Planets & Moons 7/50

Storm Giant who strikes his wife with lightning

7/51 Planets and Moons 行星和卫星
The Moons of Saturn 土星之卫星

Thrymr 索列姆

Planets & Moons 7/51

King of the frost giants killed by Thor's hammer

7/52 Planets and Moons 行星和卫星
The Moons of Saturn 土星之卫星

Aegir 埃吉尔

5 Water 水

Planets & Moons 7/52

Norse Sea God and his 9 daughters brew and drink beer

7/53 Planets and Moons 行星和卫星
The Moons of Saturn 土星之卫星

S/2007 S 3

7/54 — The Moons of Saturn 土星之卫星
Planets and Moons 行星和卫星

Bestla 贝斯特拉

5 水 Water

Planets & Moons
7/54

Norse Frost Giantess; Mother of Odin

7/55 — The Moons of Saturn 土星之卫星
Planets and Moons 行星和卫星

S/2004 S 7

7/56 — The Moons of Saturn 土星之卫星
Planets and Moons 行星和卫星

S/2006 S 3

7/57 — The Moons of Saturn 土星之卫星
Planets and Moons 行星和卫星

Chemical Elements
74

Fenrir 芬里厄

Constellation
57

Planets & Moons
7/57

Asteroids & Moons, Comets
97/1, 97/2

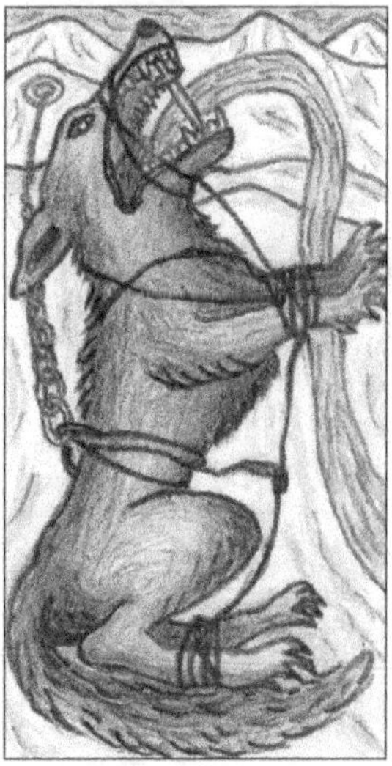

Lunar Mansion
15

Monstrous wolf bound by fetters made by elves

7/58 🌑 Planets and Moons 行星和卫星 The Moons of Saturn 土星之卫星 Surtur 苏尔特尔　　2 火 Fire  Planets & Moons 7/58 King of the fire giants; Holder of the flaming sword	**7/59** 🌑 Planets and Moons 行星和卫星 The Moons of Saturn 土星之卫星 Chemical Elements 59　Kari 卡利　　5 水 Water  Planets & Moons 7/59 Tarot ♠21, 22 Ruler of the north wind; Bringer of snow and blizzards
7/60 🌑 Planets and Moons 行星和卫星 The Moons of Saturn 土星之卫星 Ymir 尤弥尔　　5 水 Water  Planets & Moons 7/60 Asteroids & Moons, Comets 34, 46 Ymir emerges from the ice nourished by cow's milk	**7/61** 🌑 Planets and Moons 行星和卫星 The Moons of Saturn 土星之卫星 Loge (Logi) 洛格（罗吉）　　2 火 Fire  Planets & Moons 7/61 Fire Giant wins a meat eating contest

Giant; Father of Logi (Fire), Kari (Wind) and Aegir (Sea)

Planets and Moons 行星和卫星

This section includes all planets, dwarf planets and their moons as well as some potential dwarf planets and their moons between Mercury and Sedna. Relationships with days of the week, the Chinese 5 elements, the directions, the 88 Constellations, the 28 Lunar Mansions, the 12 Earthly Branches, Asteroids and Moons and Tarot cards are included.

Uranus and its Moons
天王星及其卫星

Uranus has 27 moons which are named after characters in Shakespeare's plays and the poem *Rape of the Lock* by Alexander Pope. When two different plays have characters with the same name, I have included two images. For example, Portia is a lawyer in *The Merchant of Venice* and the wife of Brutus in *Julius Caesar*. Therefore two different images, showing Portia as a Venetian lawyer and a Roman wife, are included.

8 Planets and Moons 行星和卫星

Chemical Elements 92 — Uranus 天王星 ♅ ⚷

Sky God and wife Gaea (Earth)
古希腊天神乌剌诺斯和妻子土地女神

Planets & Moons 8

8/1 Planets and Moons 行星和卫星
Moons of Uranus 天王星的卫星

Cordelia 寇蒂莉亚

King Lear

Daughter of King Lear

Planets & Moons 8/1

8/2 Planets and Moons 行星和卫星
Moons of Uranus 天王星的卫星

Ophelia 欧菲莉亚   5 Water 水

Hamlet, Prince of Denmark

Lady who loved Hamlet

Planets & Moons 8/2

---

(1) 8/3 Planets and Moons 行星和卫星
Moons of Uranus 天王星的卫星

Bianca 比恩卡

Bianca with "Language Teacher" Lucentio

Taming of the Shrew

Daughter of Baptista

Planets & Moons 8/3

(2) 8/3 Planets and Moons 行星和卫星
Moons of Uranus 天王星的卫星

Bianca 比恩卡

Planets & Moons 8/3

With Embroidered Handkerchief

Othello, the Moor of Venice

Cassio's Mistress

8/4 Planets and Moons 行星和卫星
Moons of Uranus 天王星的卫星

Cressida 克瑞西达

Planets & Moons 8/4

Troilus and Cressida

Lady loved by Troilus

8/5 Planets and Moons 行星和卫星
Moons of Uranus 天王星的卫星

Desdemona 苔丝狄蒙娜

Planets & Moons 8/5

Othello, the Moor of Venice

Wife of Othello

(1) 8/6 Planets and Moons 行星和卫星
Moons of Uranus 天王星的卫星

Juliet 朱丽叶

Planets & Moons 8/6

Romeo and Juliet

Lover and Wife of Romeo

(2) 8/6 — Planets and Moons 行星和卫星
Moons of Uranus 天王星的卫星

Juliet 朱丽叶

Planets & Moons 8/6

Measure for Measure

Lover of Claudio who is expecting his Baby

(1) 8/7 — Planets and Moons 行星和卫星
Moons of Uranus 天王星的卫星

Portia 波西亚

Planets & Moons 8/7

...as a Lawyer

The Merchant of Venice

Wife of Bassanio

(2) 8/7 — Planets and Moons 行星和卫星
Moons of Uranus 天王星的卫星

Portia 波西亚

Planets & Moons 8/7

Julius Caesar

Wife of Brutus

8/8 — Planets and Moons 行星和卫星
Moons of Uranus 天王星的卫星

Rosalind 罗斯兰

Planets & Moons 8/8

...as Ganymede

As You Like it

Daughter of exiled Duke

8/9 — Moons of Uranus 天王星的卫星

Chemical Elements 13

Cupid 丘比特

Planets & Moons 8/9

Asteroids & Moons, Comets 8, 14, 72

Timon of Athens

Boy at Timon's Banquet

8/10 — Moons of Uranus 天王星的卫星

Belinda 比琳达

Planets & Moons 8/10

Rape of the Lock

Lady who lost her Lock of Hair

8/11 — Moons of Uranus 天王星的卫星

Perdita 珀迪塔

Planets & Moons 8/11

The Winter's Tale

Daughter of King Leontes

8/12 — Moons of Uranus 天王星的卫星

Puck 帕克

3 Earth

Hobgoblin

Planets & Moons 8/12

Robin Goodfellow 又名好人罗宾

A Midsummer Night's Dream

The Mischievous Sprite

8/13

Planets and Moons 行星和卫星
Moons of Uranus
天王星的卫星

Mab
麦布女王

Air 空气 | **1** 木 Wood

Planets & Moons 8/13

Romeo and Juliet

Fairies' Midwife

8/14

Planets and Moons 行星和卫星
Moons of Uranus
天王星的卫星

Miranda
米兰达

Planets & Moons 8/14

The Tempest

Daughter of Prospero

8/15

Planets and Moons 行星和卫星
Moons of Uranus
天王星的卫星

Ariel
爱丽儿

Air 空气 | **1** 木 Wood

Planets & Moons 8/15

Rape of the Lock; The Tempest

A Sylph or Spirit of Air

8/16

Planets and Moons 行星和卫星
Moons of Uranus
天王星的卫星

Chemical Elements 27,28 | Umbriel
昂不雷尔

3 土 Earth

Planets & Moons 8/16

Rape of the Lock

A Hateful Gnome of Earth

8/17 — Moons of Uranus 天王星的卫星

Planets and Moons 行星和卫星

Titania
提泰妮娅

Air 空气 / 1 木 Wood

Planets & Moons
8/17

A Midsummer Night's Dream

Queen of the Fairies

8/18 — Moons of Uranus 天王星的卫星

Planets and Moons 行星和卫星

Oberon
奥布朗

Air 空气 / 1 木 Wood

Planets & Moons
8/18

A Midsummer Night's Dream

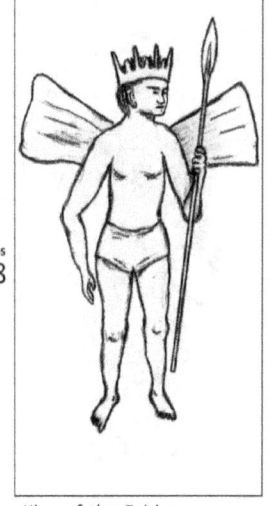

King of the Fairies

8/19 — Moons of Uranus 天王星的卫星

Planets and Moons 行星和卫星

Francisco
弗兰西斯科

Planets & Moons
8/19

The Tempest

Attendant Lord to Alonso

8/20 — Moons of Uranus 天王星的卫星

Planets and Moons 行星和卫星

Caliban
卡利班

3 土 Earth

Planets & Moons
8/20

The Tempest

The Savage and Deformed Slave

8/21 Planets and Moons 行星和卫星
Moons of Uranus 天王星的卫星

Stephano
史黛普潮

5 Water 水

Planets & Moons 8/21

The Tempest

The Drunken Butler

8/22 Planets and Moons 行星和卫星
Moons of Uranus 天王星的卫星

Trinculo
特林鸠罗

Planets & Moons 8/22

The Tempest

Jester of King Alonso

8/23 Planets and Moons 行星和卫星
Moons of Uranus 天王星的卫星

Sycorax
希克拉库斯

2 Fire 火

Planets & Moons 8/23

The Tempest

Witch and Mother of Caliban

8/24 Planets and Moons 行星和卫星
Moons of Uranus 天王星的卫星

Margaret
玛格利特

Margaret dressed as Hero at a Masked Ball

Planets & Moons 8/24

Much Ado about Nothing

Lady-in-waiting to Hero

211

8/25 — Moons of Uranus 天王星的卫星
Planets and Moons 行星和卫星

Prospero
普洛斯彼罗

Planets & Moons 8/25

The Tempest

Rightful Duke of Milan

8/26 — Moons of Uranus 天王星的卫星
Planets and Moons 行星和卫星

Setebos
塞特波斯

Planets & Moons 8/26

The Tempest

Caliban's Cruel God

8/27 — Moons of Uranus 天王星的卫星
Planets and Moons 行星和卫星

Ferdinand
腓迪南

Planets & Moons 8/27

The Tempest

Son of King Alonso

Planets and Moons 行星和卫星

This section includes all planets, dwarf planets and their moons as well as some potential dwarf planets and their moons between Mercury and Sedna. Relationships with days of the week, the Chinese 5 elements, the directions, the 88 Constellations, the 28 Lunar Mansions, the 12 Earthly Branches, Asteroids and their Moons, and Tarot cards are included.

Neptune and its Moons
海王星及其卫星

Neptune has 13 moons.

9/2 🌙 Planets and Moons 行星和卫星 **The Moons of Neptune** 海王星之卫星 Thalassa 塔拉萨　　　**5** 水 Water Planets & Moons 9/2 Wife of Sea God Pontus; Mother of Aphrodite	**9/3** 🌙 Planets and Moons 行星和卫星 **The Moons of Neptune** 海王星之卫星 Despina 德丝碧娜　　**5** 水 Water Conste-llation 80,86 Planets & Moons 9/3 Asteroids & Moons, Comets 98, 111 Tarot ♠12 ♣12 ♦12 ♥12  Lunar Mansion 25 Earthly Branch **7** 午马 Horse Horse-headed Lady; Daughter of Sea God Poseidon
9/4 🌙 Planets and Moons 行星和卫星 **The Moons of Neptune** 海王星之卫星 Galatea 伽拉忒亚　　**5** 水 Water Planets & Moons 9/4  Lady of the Milky White Foam pursued by Polyphemus	**9/5** 🌙 Planets and Moons 行星和卫星 **The Moons of Neptune** 海王星之卫星 Larissa 拉里萨　　　**5** 水 Water Planets & Moons 9/5 Thessalian Nymph with a Hydria

9/6 🌑	Planets and Moons 行星和卫星 The Moons of Neptune 海王星之卫星 S/2004 N1	**9/7** 🌑	Planets and Moons 行星和卫星 The Moons of Neptune 海王星之卫星 Proteus 普罗透斯　　5水 Water Planets & Moons 9/6 Herdsman of Poseidon's Seals
9/8 🌑	Planets and Moons 行星和卫星 The Moons of Neptune 海王星之卫星 Triton 特里同　　5水 Water Planets & Moons 9/7 Son of Sea God Poseidon	**9/9** 🌑	Planets and Moons 行星和卫星 The Moons of Neptune 海王星之卫星 Nereid 涅瑞伊得斯　　5水 Water Planets & Moons 9/8 Nymph of the Sea

9/10 Planets and Moons 行星和卫星
The Moons of Neptune 海王星之卫星
Halimede 哈利墨得 — 5 水 Water

Lady of the Brine

9/11 Planets and Moons 行星和卫星
The Moons of Neptune 海王星之卫星
Sao 圣保罗 — 5 水 Water

Lady of Rescue

9/12 Planets and Moons 行星和卫星
The Moons of Neptune 海王星之卫星
Laomedeia 拉俄墨得亚 — 5 水 Water

Lady who Leads

9/13 Planets and Moons 行星和卫星
The Moons of Neptune 海王星之卫星
Psamathe 普萨玛忒 — 5 水 Water

Lady of the Sand

Planets and Moons 行星和卫星

This section includes all planets, dwarf planets and their moons as well as some potential dwarf planets and their moons between Mercury and Sedna. Relationships with the Chinese 5 elements, the 88 Constellations, the 28 Lunar Mansions, Asteroids and Tarot cards, are included.

Orcus, Huya, Pluto, Ixion, Salacia, Varuna, Haumea, Quaoar, Makemake, Varda, 2007 OR10, Eris, Sedna and their Moons

亡神星，雨神星，冥王星，伊克西翁，萨拉喀亚，伐楼拿，妊神星，创神星，鸟神星，瓦尔达，2007 OR10，阋神星，赛德娜及其卫星

Four of these bodies (Pluto, Eris, Haumea and Makemake) have been designated as dwarf planets and the others are potential dwarf planets. A blank card represents the unnamed potential dwarf planet 2007 OR10.

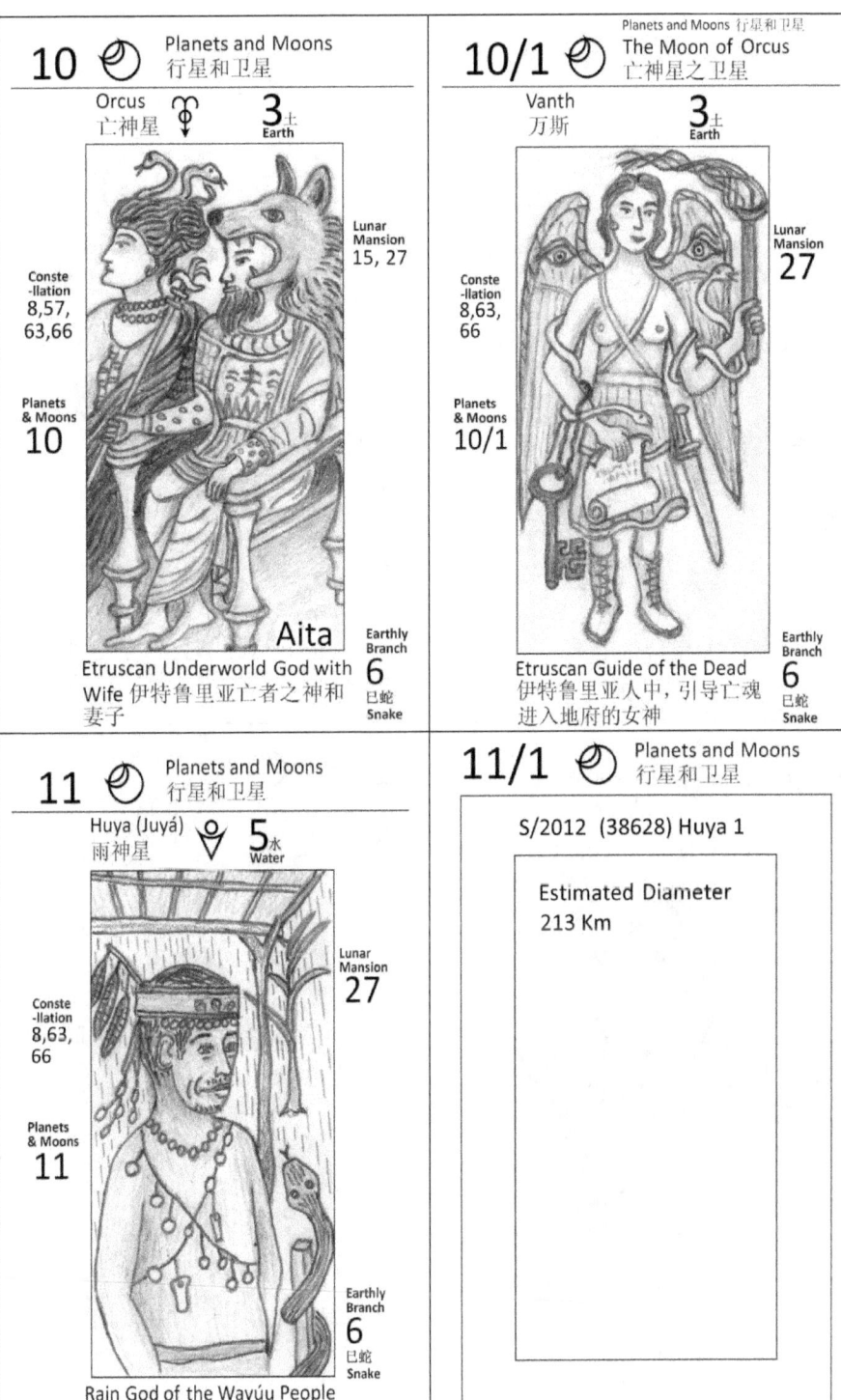

12/4 🌙 Planets and Moons 行星和卫星 The Moons of Pluto 冥王星之卫星 Kerberos **3** 土 Earth 科伯罗司 Conste-llation **25, 29, 50** Planets & Moons **12/4** Asteroids & Moons, Comets **4** Lunar Mansion **16**  Earthly Branch **11** 戌狗 Dog Guard dog of the Underworld; Hades' dog 地狱犬；哈德斯的犬	**12/5** 🌙 Planets and Moons 行星和卫星 The Moons of Pluto 冥王星之卫星 Hydra 许德拉 Conste-llation **44** Planets & Moons **12/5**  Nine-headed Serpent killed by Heracles 亦称九头蛇被赫拉克勒斯杀死
13 🌙 Planets and Moons 行星和卫星 Ixion ⛢ 伊克西翁 Conste-llation **8, 63, 66** Planets & Moons **13** Lunar Mansion **27**  Earthly Branch **6** 巳蛇 Snake Punished eternally by Zeus 被宙斯绑在一个永不停止的轮子上	**14** 🌙 Planets and Moons 行星和卫星 Salacia **5** 水 Water 萨拉喀亚 Planets & Moons **14** Asteroids & Moons, Comets **38**  Goddess of Salt Water with Sea God Husband Neptune 海之女神萨拉喀亚和丈夫尼普顿

14/1 Planets and Moons 行星和卫星
The Moon of Salacia
萨拉喀亚星之卫星

Actaea
阿克泰亚

Nymph of the Sea Shore
阿克泰亚海边仙女

15 Planets and Moons
行星和卫星

Varuna
伐楼拿   ♅   5 水 Water

Indian Sky God who became a
Sea God 印度天空之神变成
海洋之神

16 Planets and Moons
行星和卫星

Haumea
妊神星

Hawaiian Goddess of Fertility
夏威夷生育女神

16/1 Planets and Moons 行星和卫星
The Moons of Haumea
妊神星之卫星

Namaka
娜玛卡   5 水 Water

Hawaiian Sea Goddess
夏威夷海洋女神

16/2 Planets and Moons 行星和卫星
The Moons of Haumea 妊神星之卫星

Hi'iaka 希亚卡

Planets & Moons
16/2

Goddess of Hawaiian Hula Dancers 夏威夷草裙舞神

17 Planets and Moons 行星和卫星

Quaoar 创神星

Planets & Moons
17

Creation Deity of Tongva Tribe of California dances the World into Existence 加州通格瓦部落创世之神夸欧尔，通过舞蹈和旋转创造万物

17/1 Planets and Moons 行星和卫星
The Moon of Quaoar 创神星之卫星

Weywot 维沃特

Planets & Moons
17/1

Weywot dances the Earth Goddess Chehooit into Existence
维沃特天神通过舞蹈创造大地女神切胡伊特
Dancing Sky God of the Tongva People 通格瓦人的舞蹈天神

18 Planets and Moons 行星和卫星

Makemake 鸟神星

Planets & Moons
18

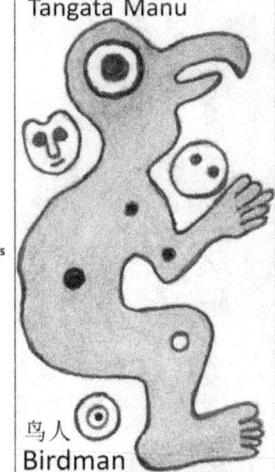

Tangata Manu

鸟人 **Birdman**

Creator and Fertility God of Rapa Nui (Easter Island) 马奇马奇是复活节岛拉帕努伊人中人类创造者与生殖之神

18/1 Planets and Moons 行星和卫星

S/2015 (136472) Makemake 1

19 Planets and Moons 行星和卫星

Varda 瓦尔达

2 火 Fire

Planets & Moons
19

Asteroids & Moons, Comets
25, 39

门 17

Creator of the Stars and Queen of the Valar in Tolkien's *Legendarium* 瓦尔达在托尔金的传说故事集里是众维拉中的女王，星辰创造者。

19/1 Planets and Moons 行星和卫星
The Moon of Varda 瓦尔达 星之卫星

Ilmarë
伊尔玛瑞

Planets & Moons
19/1

Asteroids & Moons, Comets
25, 39

门 17

Handmaiden of Varda and Chief of the Maiar in Tolkien's *Legendarium*
伊尔玛瑞是瓦尔达的侍女。

20 Planets and Moons 行星和卫星

2007 OR 10

21 Planets and Moons 行星和卫星

Eris 阋神星 — Air 空气 — 1 Wood 木

Goddess of Discord and Strife
厄里斯，纷争兼不和女神

21/1 Planets and Moons 行星和卫星 — The Moon of Eris 阋神星之卫星

Chemical Elements 85 — Dysnomia 迪斯诺美亚 — 2 Fire 火

Spirit of Lawlessness,
Daughter of Eris 违法女神,
厄里斯的女儿

22 Planets and Moons 行星和卫星

Sedna 赛德娜 — 5 Water 水

Inuit Sea Goddess, Mother of all
Sea Creatures 因努伊特人的海
洋女神,是一切海洋生物的母亲

Asteroids and Moons, Comets 小行星及其卫星和彗星

This is a selection of 120 asteroids (known also as Minor Planets) most of which have a diameter of more than 100 kilometres. Some small inner solar system objects and near earth objects are included to balance the picture but most are members of the main asteroid belt. There are also Trojan asteroids, plus a few extremely distant trans-Neptunian objects. One short-period comet, one medium-period comet and two long-period comets have been added to this list making a total of 124 cards plus extra cards for asteroid moons. Unnamed asteroid moons are represented by blank cards.

3 Asteroids and Moons, Comets 小行星及其卫星和彗星	**4** Asteroids and Moons, Comets 小行星及其卫星和彗星
1566 Icarus 伊卡洛斯 **2** 火 Fire  Asteroids & Moons, Comets **3** Icarus falls into the Aegean Sea when the sun melts his wings 伊卡洛斯飞太阳太近使蜡翼融化坠入爱琴海身亡	1865 Cerberus 刻耳柏洛斯 **3** 土 Earth  Conste-llation 25,29, 50 Planets & Moons 12/4 Asteroids & Moons, Comets **4** Lunar Mansion **16** Guard dog of the Underworld; Hades' dog 地狱犬：哈德斯的犬 Earthly Branch **11** 戌狗 Dog
5 Asteroids and Moons, Comets 小行星及其卫星和彗星	**6** Asteroids and Moons, Comets 小行星及其卫星和彗星
1620 Geographos 地理星 **3** 土 Earth  Asteroids & Moons, Comets **5** The Geographer 地理学家 Strabo 斯特拉波 (64/63 BC – ca. 24 AD)	Chemical Elements **73** 2102 Tantalus 坦塔洛斯 **5** 水 Water 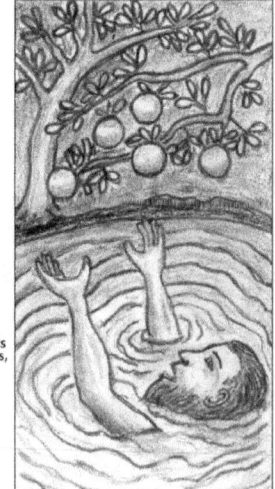 Asteroids & Moons, Comets **6** Tantalus' punishment: Eternal Hunger and Thirst 坦塔洛斯的惩罚：永远忍受饥渴的折磨

7	Asteroids and Moons, Comets 小行星及其卫星和彗星

Chemical Elements 63 · **1685 Toro** 托罗 · **3** Earth

Constellation 16

Lunar Mansion 9

Asteroids & Moons, Comets 7

Zodiac Sign 2 Taurus

Earthly Branch 2 丑牛 Ox

Ancient cult of the bull continues in modern Spain 古代崇拜牛在当今西班牙仍然继续。

8	Asteroids and Moons, Comets 小行星及其卫星和彗星

Chemical Elements 13 · **433 Eros** 爱神星

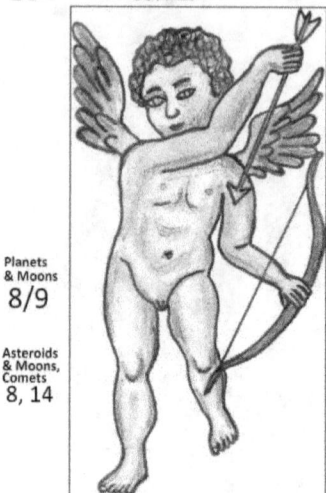

Planets & Moons 8/9

Asteroids & Moons, Comets 8, 14

Ancient Greek God of Love and Fertility 古希腊的爱神和生育神厄洛斯

9	Asteroids and Moons, Comets 小行星及其卫星和彗星

1864 Daedalus 代达罗斯 · **1** Wood

Asteroids & Moons, Comets 9

Daedalus makes a wooden cow for Pasiphaë the mother of the Minotaur 代达罗斯为弥诺陶洛斯的母亲帕西法厄做了一个木牛

10	Asteroids and Moons, Comets 小行星及其卫星和彗星

Chemical Elements 2, 42, 60, 89, 91 · **1862 Apollo** 阿波罗 · **1** Wood · **Day** 7 Sun

Shang Suns 3

Planets & Moons 0, 6/66

Asteroids & Moons, Comets 10

凩 19

Apollo, Greek God of Music, Prophecy, Medicine, Higher Learning, Archery and the Sun 阿波罗是希腊音乐神，预言家，药王，学者，太阳神及箭神。

10/1 Asteroids and Moons, Comets 小行星及其卫星和彗星 S/2005 (1862)1	**11** Asteroids and Moons, Comets 小行星及其卫星和彗星 1981 Midas 弥达斯 5 水 Water Planets & Moons **6/46** Asteroids & Moons, Comets **11**  Midas captures the satyr Seilenos at a fountain in his garden 弥达斯在自家花园的喷泉边上捉到了萨梯西勒诺斯
12 Asteroids and Moons, Comets 小行星及其卫星和彗星 1627 Ivar 伊瓦尔 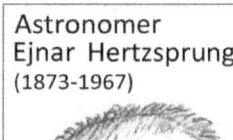 Astronomer Ejnar Hertzsprung (1873-1967) Asteroids & Moons, Comets **12**  Ejnar named this asteroid after his brother Ivar 丹麦天文学家埃希纳·赫茨普龙以其兄弟伊瓦尔命名此星	**13** Asteroids and Moons, Comets 小行星及其卫星和彗星 2101 Adonis 阿多尼斯 Lunar Mansion **13** Asteroids & Moons, Comets **13**  God of Crops and Fertility who was killed by a wild boar 植物神和生育神，被野猪撞伤致死 Earthly Branch **12** 亥猪 Pig

14	Asteroids and Moons, Comets 小行星及其卫星和彗星		15	Asteroids and Moons, Comets 小行星及其卫星和彗星

Chemical Elements 13 — 1221 **Amor** 阿莫尔

Conste-llation 58, 67 — 434 **Hungaria** 匈牙利 — 3 Earth 土

Planets & Moons 8/9

Asteroids & Moons, Comets 8, 14

Roman Boy-god of Love 古罗马爱神

Asteroids & Moons, Comets 15

Hungary Magyarország 匈牙利国徽

16	Asteroids and Moons, Comets 小行星及其卫星和彗星		17	Asteroids and Moons, Comets 小行星及其卫星和彗星

Chemical Elements 98 — 341 **California** 加利福尼亚星 — 3 Earth 土

Chemical Elements 17 — 8 **Flora** 花神星 — 1 Wood 木

Eureka

Conste-llation 42, 54

Planets & Moons 6/8, 6/24, 7/25

Asteroids & Moons, Comets 16

California 加利福尼亚州州徽

Asteroids & Moons, Comets 17, 54

Roman Goddess of Flowering Plants 罗马花之女神

(1) 18	Asteroids and Moons, Comets 小行星及其卫星和彗星

2P/ Encke (Comet) 恩克彗星 2 火 Fire

Asteroids & Moons, Comets 18

Comet named after German Astronomer Johann Franz Encke (1791-1865) 德国天文学家约翰·弗朗茨·恩克

---

(2) 18	Asteroids and Moons, Comets 小行星及其卫星和彗星

2P/ Encke (Comet) 恩克彗星 2 火 Fire

Asteroids & Moons, Comets 18

Comet named after German Astronomer Johann Franz Encke (1791-1865) 德国天文学家约翰·弗朗茨·恩克

19	Asteroids and Moons, Comets 小行星及其卫星和彗星

40 Harmonia 谐神星

Asteroids & Moons, Comets 19, 51

Greek Goddess of Harmony and Concord 哈耳摩尼亚是古希腊和谐女神

20	Asteroids and Moons, Comets 小行星及其卫星和彗星

18 Melpomene ✝ ✱

司曲星

Asteroids & Moons, Comets 20

Muse of Tragedy 墨尔波墨涅是悲剧女神，文艺九女神缪斯之一

21 Asteroids and Moons, Comets 小行星及其卫星和彗星		**22** Asteroids and Moons, Comets 小行星及其卫星和彗星	

12 Victoria 凯神星

Asteroids & Moons, Comets
21

Roman Goddess of Victory
罗马胜利女神

27 Euterpe 司箫星

Asteroids & Moons, Comets
22

Muse of Flute Playing
欧忒耳珀是九缪斯之一，演奏双管长笛

23 Asteroids and Moons, Comets 小行星及其卫星和彗星		**24** Asteroids and Moons, Comets 小行星及其卫星和彗星	

4 Vesta 灶神星 2 Fire

Asteroids & Moons, Comets
23, 36

Roman Goddess of the Hearth
维斯塔是罗马女灶神

51 Nemausa 禽神星 3 Earth

Asteroids & Moons, Comets
24

Nimes, France. Roman colony of Nemausa 法国尼姆在古罗马殖民时期的名字是 Nemausa

25 Asteroids and Moons, Comets 小行星及其卫星和彗星 Chemical Elements **92** **30** Urania 司天星 Planets & Moons **19,19/1** Asteroids & Moons, Comets **25** 阁 17  Muse of Astronomy 乌拉尼亚是掌管天文的缪斯	**26** Asteroids and Moons, Comets 小行星及其卫星和彗星 Chemical Elements **77** **7** Iris 虹神星 Asteroids & Moons, Comets **26**  Goddess of the Rainbow 古希腊彩虹女神
27 Asteroids and Moons, Comets 小行星及其卫星和彗星 **9** Metis 颖神星 Planets & Moons **6/1** Asteroids & Moons, Comets **27** 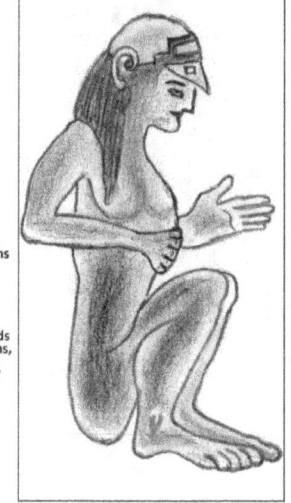 First Wife of Zeus 墨提斯是宙斯的第一妻子	**28** Asteroids and Moons, Comets 小行星及其卫星和彗星 **63** Ausonia 澳女星 **3** 土 Earth Conste-llation **58, 67** Asteroids & Moons, Comets **28** 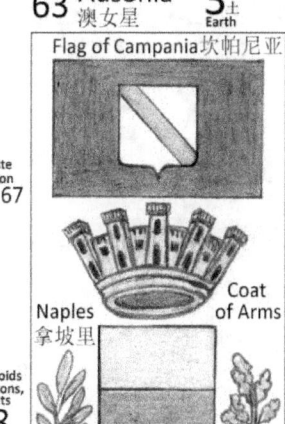 Flag of Campania 坎帕尼亚 Naples 拿坡里 Coat of Arms Ausonia: ancient name for Campania in Italy 奥索尼亚:是坎帕尼亚在古意大利时期的名字

29 Asteroids and Moons, Comets 小行星及其卫星和彗星 **192** Nausikaa 瑙西卡  Nausicaa meets the shipwrecked Odysseus 奥德修斯海难后与瑙西卡相遇	**30** Asteroids and Moons, Comets 小行星及其卫星和彗星 **20** Massalia 王后星  Massalia: ancient Greek name for Marseille, France 马赛市徽，马萨利亚是法国马赛市在古希腊时期名字
31 Asteroids and Moons, Comets 小行星及其卫星和彗星 **6** Hebe 韶神星　**5** Water 水  Cup-bearer of the gods. Goddess of youth. 赫伯是青春女神，为神酌酒的美少女	**32** Asteroids and Moons, Comets 小行星及其卫星和彗星 Chemical Elements **71** **21** Lutetia 司琴星　**5** Water 水  Lutetia: Roman name for Paris, France 鲁特西亚：古罗马时期法国首都巴黎的名字

33	Asteroids and Moons, Comets 小行星及其卫星及彗星

19 Fortuna
命神星

Asteroids & Moons, Comets
33

Roman Goddess of Fortune
福尔图娜：古罗马命运女神

34	Asteroids and Moons, Comets 小行星及其卫星和彗星

42 Isis
育神星

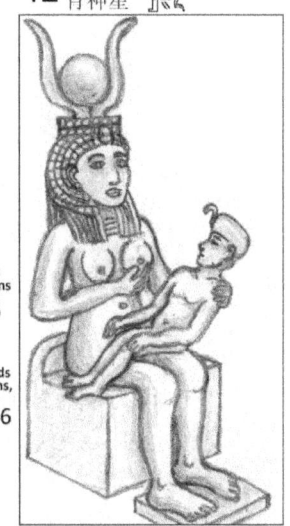

Planets & Moons
6/5

Asteroids & Moons, Comets
34, 46

Egyptian Goddess of Protection, Healing and Fertility 古埃及生育神，守护神及医疗之神

35	Asteroids and Moons, Comets 小行星及其卫星和彗星

11 Parthenope
海妖星 5 Water

Asteroids & Moons, Comets
35

The Siren who became the Goddess of Naples 帕尔特诺佩：海妖塞壬之一，意大利城市拿坡里的女神

36	Asteroids and Moons, Comets 小行星及其卫星和彗星

46 Hestia
司祭星 2 Fire

Asteroids & Moons, Comets
23, 36

Greek Goddess of the Hearth
赫斯提亚是希腊女灶神

37 Asteroids and Moons, Comets 小行星及其卫星和彗星	38 Asteroids and Moons, Comets 小行星及其卫星和彗星
89 Julia 淫神星  Asteroids & Moons, Comets **37** Saint Julia of Corsica, a Carthaginian Saint 圣朱莉娅是科西嘉岛和古迦太基国的圣人	**29 Amphitrite** 海后星 **5 Water** ☾* Planets & Moons **14**  Asteroids & Moons, Comets **38** Amphitrite with Sea-god husband Poseidon 安菲特里忒和海神丈夫波塞冬
39 Asteroids and Moons, Comets 小行星及其卫星和彗星	40 Asteroids and Moons, Comets 小行星及其卫星和彗星
5 Astraea 义神星 ⛎ ♎ **2 Fire** Conste-llation **51** Planets & Moons **19, 19/1** Asteroids & Moons, Comets **39**  ♐17 Astraea: Star Maiden. Dike: Goddess of Justice 正义女神狄克飞天后变成少女星爱斯翠雅	**13 Egeria** 芙女星 ⚵ **5 Water**  Asteroids & Moons, Comets **40** Water Nymph and Wife of Numa Pompilius the 2nd king of Rome 埃杰里亚是水中女仙，第二代罗马国王努马·庞皮利乌斯的妻子

235

41	Asteroids and Moons, Comets 小行星及其卫星和彗星	**42**	Asteroids and Moons, Comets 小行星及其卫星和彗星

409 Aspasia 阿斯帕齐娅

Asteroids & Moons, Comets
41

Famous Courtesan and Wife of Pericles 阿斯帕齐娅是古希腊聪明而有权势的女人，雅典政治家佩里克勒斯的妻子

14 Irene 司宁星

Asteroids & Moons, Comets
42

Eirene, Goddess of Peace, holding infant Ploutos (Wealth) 埃瑞涅是和平女神，抱着婴儿财神普路托斯

43	Asteroids and Moons, Comets 小行星及其卫星和彗星	**44**	Asteroids and Moons, Comets 小行星及其卫星和彗星

91 Aegina 河神星   5 Water

Asteroids & Moons, Comets
43

River Nymph abducted by Zeus 女河神埃吉娜被宙斯拐走

53 Kalypso 岛神星   5 Water

Planets & Moons
7/17

Asteroids & Moons, Comets
44

Calypso finds Odysseus on the seashore 卡利普索在岸边发现了奥德修斯

45 Asteroids and Moons, Comets 小行星及其卫星和彗星

15 Eunomia 司法星 **1 木 Wood**

Asteroids & Moons, Comets
45

Goddess of Good Government, Law and Order 欧诺弥亚是法律和秩序的女神

46 Asteroids and Moons, Comets 小行星及其卫星和彗星

85 Io 犊神星 **5 水 Water**

Planets & Moons
6/5

Asteroids & Moons, Comets
34, 46

Lover of Zeus changed into a cow guarded by Argos
宙斯为救情人伊俄，把她变为一只牛，赫拉令长着百只眼睛的阿刚斯看守她

47 Asteroids and Moons, Comets 小行星及其卫星和彗星

Chemical Elements
23

240 Vanadis 凡娜迪丝

Day **5** Fri

Lunar Mansion
13

Asteroids & Moons, Comets
47

Earthly Branch
12 亥猪 Pig

Vanadis (Freya) Norse Goddess of Love with "Battle Swine" 也叫弗蕾亚，是北欧神话中爱与美的女神，曾骑野猪参与战斗

48 Asteroids and Moons, Comets 小行星及其卫星和彗星

3 Juno 婚神星

Conste-llation
72

Asteroids & Moons, Comets
48, 52

Roman Goddess of Women and Marriage; Wife of Jupiter
朱诺是罗马婚姻保护神，主神朱庇特的妻子

49	Asteroids and Moons, Comets 小行星及其卫星和彗星	50	Asteroids and Moons, Comets 小行星及其卫星和彗星

324 Bamberga 斑贝格星 3 Earth 土

Conste-llation 69, 74

Planets & Moons 6, 6/7, 7/5

Asteroids & Comets 49

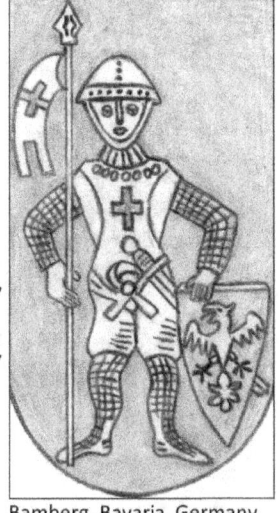

Bamberg, Bavaria, Germany
德国巴伐利亚州斑贝格市市徽

34 Circe 巫神星

Conste-llation 37, 40

Asteroids & Moons, Comets 50

Zodiac Sign 5 Leo

The witch Circe changes Odysseus' men into animals
女巫喀耳刻将奥德修斯船上的人变成了动物

51	Asteroids and Moons, Comets 小行星及其卫星和彗星	52	Asteroids and Moons, Comets 小行星及其卫星和彗星

58 Concordia 协神星 5 Water 水

Asteroids & Moons, Comets 19, 51

Roman Goddess of Harmony and Agreement
康考迪亚是罗马和谐女神

103 Hera 后神星

Asteroids & Moons, Comets 48, 52

Queen of the Gods; Wife of Zeus; Goddess of Marriage
赫拉是天后，宙斯妻子，婚姻保护神

53 Asteroids and Moons, Comets 小行星及其卫星和彗星

45 Eugenia
号香女星

Asteroids & Moons, Comets
53

Empress Eugenie of France (1826-1920)
法国王后尤金尼亚

53/1 Asteroids and Moons, Comets 小行星及其卫星和彗星

S/2004(45)1

53/2 Asteroids and Moons, Comets 小行星及其卫星和彗星

Petit-Prince
小王子

Asteroids & Moons, Comets
53/2

The Little Prince Louis Napoleon at the age of 14
14 岁的小王子路易·拿破仑

54 Asteroids and Moons, Comets 小行星及其卫星和彗星

Chemical Elements 17 — 410 Chloris 克洛莉丝 — 1 木 Wood

Asteroids & Moons, Comets
17, 54

Greek Nymph of Flowers and New Plant Growth
克洛莉丝是希腊花神和林中仙女

55	Asteroids and Moons, Comets 小行星及其卫星和彗星

1021 Flammario 弗拉马利翁

Asteroids & Moons, Comets
55, 96

French Astronomer N. Camille Flammarion (1842-1925)
法国天文学家卡米伊·弗拉马利翁

56	Asteroids and Moons, Comets 小行星及其卫星和彗星

38 Leda 卵神星

Conste-llation
76

Planets & Moons
6/10

Asteroids & Moons, Comets
56

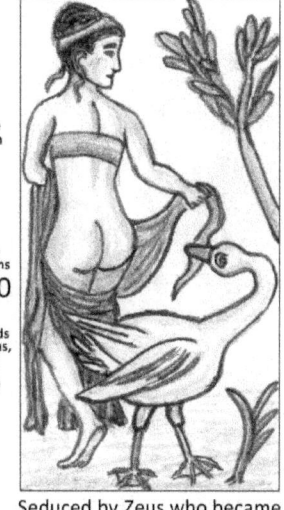

Seduced by Zeus who became a swan
宙斯化作天鹅来探访丽妲

57	Asteroids and Moons, Comets 小行星及其卫星和彗星

128 Nemesis 罚神星 4 金 Metal

Asteroids & Moons, Comets
57

Spirit of Divine Retribution
涅墨西斯是复仇女神

58	Asteroids and Moons, Comets 小行星及其卫星和彗星

Chemical Elements
47

93 Minerva 慧神星 4 金 Metal

Asteroids & Moons, Comets
58, 66

Virgin Goddess of Creative Arts and War 弥涅耳瓦是智慧和战争女神，艺术家的保护神

58/1 Asteroids and Moons, Comets 小行星及其卫星和彗星

Gorgoneion
戈耳工之首

Conste-llations **12**

Asteroids & Moons, Comets **58/1**

Head of the Gorgon Medusa which can turn enemies to stone
蛇发女妖美杜莎之首能把敌人变成石头。

58/2 Asteroids and Moons, Comets 小行星及其卫星和彗星

Aegis
埃癸斯

Conste-llations **12**

Asteroids & Moons, Comets **58/2**

Minerva (Athena) wears an Aegis displaying the head of the Gorgon Medusa
密涅瓦（雅典娜）穿着有蛇发女妖美杜莎的埃癸斯胸甲。

59 Asteroids and Moons, Comets 小行星及其卫星和彗星

Chemical Elements **41**

71 Niobe
石女星

Asteroids & Moons, Comets **59**

10 of her 12 children were killed by Apollo and Artemis
尼奥比因自夸而被阿波罗和阿耳特弥斯杀死 10 个孩子

60 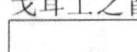 Asteroids and Moons, Comets 小行星及其卫星和彗星

55 Pandora
祸神星

Planets & Moons **7/6**

Asteroids & Moons, Comets **60**

Pandora opens the jar of evils
潘多拉打开魔盒

61	Asteroids and Moons, Comets 小行星及其卫星和彗星

41 Daphne 桂神星 1 Wood 木

Asteroids & Moons, Comets 61

Virgin Nymph who was changed into a Laurel Tree 处女达芙妮被变成了月桂树

61/1	Asteroids and Moons, Comets 小行星及其卫星和彗星

S/2008(41)1

62	Asteroids and Moons, Comets 小行星及其卫星和彗星

88 Thisbe 尽女星 3 Earth 土

Conste-llation 37, 40

Asteroids & Moons, Comets 62

Zodiac Sign 5 Leo

Thisbe hides in a cave from a lion 提斯柏为躲避狮子藏身洞穴

63	Asteroids and Moons, Comets 小行星及其卫星和彗星

39 Laetitia 喜神星

Asteroids & Moons, Comets 63

Roman Goddess of Joy and Gladness 莱堤西亚是罗马欢乐女神

64

Asteroids and Moons, Comets 小行星及其卫星和彗星

Chemical Elements: 31

148 **Gallia** 高卢星

3 Earth 土

Asteroids & Moons, Comets: 64

Liberté, égalité, fraternité

Gallia (Gaul) was the Roman name for France 法国国徽，佳利雅是古罗马时期法国的名字

65

Asteroids and Moons, Comets 小行星及其卫星和彗星

532 **Herculina** 大力神星

Constellation: 63, 65

Planets & Moons: 6/34

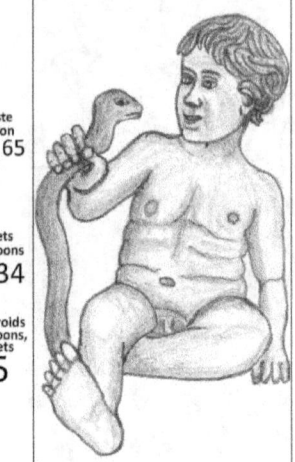

Asteroids & Moons, Comets: 65

Lunar Mansion: 27

8 month old Hercules strangles a snake 赫拉克勒斯八个月大时扼死一条蛇

Earthly Branch: 6 巳蛇 Snake

66

Asteroids and Moons, Comets 小行星及其卫星和彗星

Chemical Elements: 46

2 **Pallas** 智神星

4 Metal 金

Constellation: 63

Asteroids & Moons, Comets: 58, 66

Lunar Mansion: 27

Pallas Athena Goddess of Wisdom 帕拉斯·雅典娜是智慧女神

Earthly Branch: 6 巳蛇 Snake

67

Asteroids and Moons, Comets 小行星及其卫星和彗星

216 **Kleopatra** 艳后星

Asteroids & Moons, Comets: 67

Cleopatra VII Philopator (69-30 BC) Queen of Egypt 克娄巴特拉七世是埃及女王

67/1 Asteroids and Moons, Comets 小行星及其卫星和彗星	67/2 Asteroids and Moons, Comets 小行星及其卫星和彗星
Cleoselene 克娄巴特拉·塞勒涅  Cleopatra Selene Queen of Mauretania 克娄巴特拉·塞勒涅是毛里塔尼亚王后，埃及艳后的女儿	**Alexhelios** 亚历山大·赫利俄斯  Alexander Helios, son of Cleopatra, born 40 BC 亚历山大·赫利俄斯是埃及艳后的儿子，出生于公元前40年
68 Asteroids and Moons, Comets 小行星及其卫星和彗星	69 Asteroids and Moons, Comets 小行星及其卫星和彗星
804 **Hispania** 伊斯帕尼亚 3 Earth  PLUS ULTRA: "Further Beyond" Hispania: Latin name for Spain — España 伊斯帕尼亚是西班牙在罗马时期的名字	243 **Ida** 义女星 3 Earth  Ida nurses baby Zeus in a cave on Mount Ida in Crete 艾达在克里特岛艾达山的山洞里给婴儿宙斯喂奶

69/1 — Asteroids and Moons, Comets 小行星及其卫星和彗星	70 — Asteroids and Moons, Comets 小行星及其卫星和彗星
Dactyl 达克堤利 3 ⊕ Earth  Rhea presses her fingers into the earth giving birth to the Dactyls 瑞亚将手指插入泥土里，于是从地里生出了达克堤利（手指）	**386 Siegena** 锡根市 3 ⊕ Earth  Conste-llation 37, 40  Zodiac Sign 5 Leo Latin name of Siegen in Nordrhein-Westfalen, Germany 德国城市锡根市市徽
71 — Asteroids and Moons, Comets 小行星及其卫星和彗星	71/1 — Asteroids and Moons, Comets 小行星及其卫星和彗星
22 Kalliope 司赋星  Calliope Muse of Epic Poetry 卡利俄佩是希腊神话中掌管史诗的缪斯	**Linus** 利诺斯  Famous Musician and son of Calliope 利诺斯是著名的音乐家，卡利俄佩的儿子

72 Asteroids and Moons, Comets 小行星及其卫星和彗星 Chemical Elements **13** **16** Psyche ☀ 灵神星 Planets & Moons **8/9** Asteroids & Moons, Comets **72**  Princess who married Cupid and became immortal 公主赛克与丘比特结婚后成为神仙	**73** Asteroids and Moons, Comets 小行星及其卫星和彗星 **349** Dembowska 邓鲍斯基 Asteroids & Moons, Comets **73**  Baron Ercole Dembowski makes 20,000 astronomical observations 男爵艾尔科雷・邓鲍斯基制作了二万个天文观测台
74 Asteroids and Moons, Comets 小行星及其卫星和彗星 **747** Winchester **3** Earth 温彻斯特 Asteroids & Moons, Comets **74**  Colonel William P. Winchester paid $3,000 to build Town Hall 上校威廉・温彻斯特出资三千美元建立了美国马萨诸塞州温彻斯特市的市政厅	**75** Asteroids and Moons, Comets 小行星及其卫星和彗星 **283** Emma 艾玛 Asteroids & Moons, Comets **75**  An unknown woman 艾玛个人资料不详

75/1 — Asteroids and Moons, Comets 小行星及其卫星和彗星

S/2003(283)1

76 — Asteroids and Moons, Comets 小行星及其卫星和彗星

Chemical Elements 32 | 241 Germania 日耳曼尼亚 | 3 Earth

Constellation 69, 74

Planets & Moons 6, 6/7, 7/5

Asteroids & Moons, Comets 76

Bundesadler

Roman name for Germany
德国国徽，德国在古罗马时期叫日耳曼尼亚

77 — Asteroids and Moons, Comets 小行星及其卫星和彗星

451 Patientia 佩兴提亚 3 Earth

Asteroids & Moons, Comets 77

Patience: a humble woman enduring adversity 一个谦逊的，能忍受困苦的女人

78 — Asteroids and Moons, Comets 小行星及其卫星和彗星

423 Diotima 狄欧蒂玛

Asteroids & Moons, Comets 78

Priestess from Mantinea who taught Socrates philosophy
来自曼提尼亚的女祭司狄欧蒂玛教授了苏格拉底的哲学思想

(1) 79 Asteroids and Moons, Comets 小行星及其卫星和彗星	(2) 79 Asteroids and Moons, Comets 小行星及其卫星和彗星
704 Interamnia 因泰拉姆尼亚 3± Earth	704 Interamnia 因泰拉姆尼亚 3± Earth
Conste-llation 58, 67  Asteroids & Moons, Comets 79	Conste-llation 58, 67  Asteroids & Moons, Comets 79
Interamnia Praetuttorium: Roman name for Teramo, Italy 泰拉莫在意大利古罗马时期的名字	Interamna Nahars: Roman name for Terni, Umbria, Italy 特尔尼在意大利古罗马时期的名字
80 Asteroids and Moons, Comets 小行星及其卫星和彗星	81 Asteroids and Moons, Comets 小行星及其卫星和彗星
Chemical Elements 63 52 Europa 欧女星 5 Water 水	130 Elektra 怂女星 3± Earth
Planets & Moons 6/6  Asteroids & Moons, Comets 80	 Asteroids & Moons, Comets 81
Zeus becomes a bull and carries Europa across the sea to Crete 宙斯化成一头白牛驮着腓尼基公主欧罗巴越海来到克里特岛	Electra sits at the tomb of her father Agamemnon 厄勒克特拉坐在父亲阿伽门农的坟墓前

81/1 Asteroids and Moons, Comets 小行星及其卫星和彗星

S/2014 (130)1

81/2 Asteroids and Moons, Comets 小行星及其卫星和彗星

S/2003 (130)1

82 Asteroids and Moons, Comets 小行星及其卫星和彗星

375 Ursula 厄休拉

Conste-llation **54**

Asteroids & Moons, Comets **82**

Saint Ursula from Dumnonia (Cornwall) 圣女厄休拉来自多姆尼亚（今康沃尔郡）

83 Asteroids and Moons, Comets 小行星及其卫星和彗星

165 Loreley 水妖星　**5** 水 Water

Asteroids & Moons, Comets **83**

The siren's singing lures boatmen to their death 水妖罗蕾莱的歌声吸引水手倾听失神触礁身亡

84 — Asteroids and Moons, Comets 小行星及其卫星和彗星	**85** — Asteroids and Moons, Comets 小行星及其卫星和彗星
24 **Themis** 司理星 Conste-llation **56** Asteroids & Moons, Comets **84** Tarot ◊6  Goddess of Justice 特弥斯是正义女神	10 **Hygiea** 健神星 Conste-llation **63, 66** Planets & Moons **6/34** Asteroids & Moons, Comets **85**   Lunar Mansion **27** Goddess of Health 海净是健康女神 Earthly Branch **6** 巳蛇 Snake
86 — Asteroids and Moons, Comets 小行星及其卫星和彗星	**87** — Asteroids and Moons, Comets 小行星及其卫星和彗星
31 **Euphrosyne** 丽神星  Asteroids & Moons, Comets **86** Goddess of Joy; one of the Graces or Charites 欧佛洛绪涅是欢乐女神，美惠三女神之一	250 **Bettina** 贝蒂娜  Asteroids & Moons, Comets **87** Baroness Bettina Caroline von Rothschild and husband 女公爵贝蒂娜·卡洛琳·冯·罗斯切尔德和她的丈夫

| 88 Asteroids and Moons, Comets 小行星及其卫星和彗星 | 88/1 Asteroids and Moons, Comets 小行星及其卫星和彗星 |

762 **Pulcova** 普尔科沃

Asteroids & Moons, Comets
88

Pulkovo Astronomical Observatory, Saint Petersburg 普尔科沃天文台位于圣彼得堡

S/2000(762)1

| 89 Asteroids and Moons, Comets 小行星及其卫星和彗星 | 90 Asteroids and Moons, Comets 小行星及其卫星和彗星 |

511 **Davida** 戴维

Asteroids & Moons, Comets
89

American Astronomer David Peck Todd (1855-1939) 美国天文学家戴维·派克·托德

106 **Dione** 坤神星

Planets & Moons
7/18

Asteroids & Moons, Comets
90

Presiding Goddess at Oracular Shrine of Dodona 主持女神狄俄涅在希腊圣地多多娜

91	Asteroids and Moons, Comets 小行星及其卫星和彗星	92	Asteroids and Moons, Comets 小行星及其卫星和彗星

92 Undina 波女星 **5** 水 Water

Asteroids & Moons, Comets
91

Water Spirit Undina emerges from a fountain 水女神温蒂妮出现在温泉中

702 Alauda 云雀

Asteroids & Moons, Comets
92

Alauda: Latin name for the Lark
Alauda：云雀的拉丁名字

92/1	Asteroids and Moons, Comets 小行星及其卫星和彗星	93	Asteroids and Moons, Comets 小行星及其卫星和彗星

Pichi üñëm 小鸟

Asteroids & Moons, Comets
92/1

Pichi üñëm "Little Bird" in the Mapuche language of Chile
在智利马普切语中是小鸟的意思

Chemical Elements
2, 42, 60, 89, 91

895 Helio 赫利奥 **2** 火 Fire **Day 7** Sun

Shang Suns
3

Planets & Moons
0, 6/66

Asteroids & Moons, Comets
1, 93

卐19

Helio: accusative case of the Greek Helios "The Sun"
Helio 是由希腊语 Helios 派生出来的，意为"太阳"

94 Asteroids and Moons, Comets 小行星及其卫星和彗星	**95** Asteroids and Moons, Comets 小行星及其卫星和彗星
65 **Cybele** 原神星　3 ♁ Earth Conste-llation 37, 40 Planets & Moons 7/21 Asteroids & Moons, Comets 94 Zodiac Sign 5 Leo  The Great Mother Goddess of Phrygia 库柏勒是古代佛里吉亚国的母亲女神	121 **Hermione** 赫女星  Asteroids & Moons, Comets 95 Daughter of Helen and Menelaus 赫耳弥俄涅是墨涅拉俄斯和海伦之女
95/1 Asteroids and Moons, Comets 小行星及其卫星和彗星	**96** Asteroids and Moons, Comets 小行星及其卫星和彗星
S/2002(121)1	107 **Camilla** 卡米伊  Asteroids & Moons, Comets 55, 96 N. Camille Flammarion (1842-1925) French Astronomer 法国天文学家卡米伊·弗拉马利翁

96/1 Asteroids and Moons, Comets 小行星及其卫星和彗星	96/2 Asteroids and Moons, Comets 小行星及其卫星和彗星
S/2016 (107) 1	S/2001 (107) 1

97 Asteroids and Moons, Comets 小行星及其卫星和彗星	97/1 Asteroids and Moons, Comets 小行星及其卫星和彗星
87 Sylvia 林神星 1 Wood  Asteroids & Moons, Comets 97 Mars finds Rhea Sylvia in a forest 战神马尔斯在森林中遇见雷亚·西尔维亚	Chemical Elements 74 Remus 雷穆斯 Lunar Mansion 15 Constellations 57 Planets & Moons 6/54, 7/29, 34, 39, 40, 46, 48, 57 Asteroids & Moons, Comets 97/1  Remus cared for by a she-wolf beside the Tiber 雷穆斯在台伯河边由母狼抚育

97/2 Asteroids and Moons, Comets 小行星及其卫星和彗星

Chemical Elements 74

Romulus
罗慕路斯

Conste-llations 57

Planets & Moons 6/54, 7/29, 34, 39, 40, 46, 48, 57

Asteroids & Moons, Comets 97/2

Lunar Mansion 15

Romulus suckled by a she-wolf
母狼给罗慕路斯喂奶

---

98 Asteroids and Moons, Comets 小行星及其卫星和彗星

153 Hilda
女武神

Conste-llation 80

Planets & Moons 6/39, 7/20, 9/3

Asteroids & Moons, Comets 98

Lunar Mansion 25

Earthly Branch 7 年马 Horse

Valkyrie or Battle Maiden Hilda rides through the sky
北欧神话中女武神希露德骑马飞过天空

---

99 Asteroids and Moons, Comets 小行星及其卫星和彗星

Chemical Elements 69

279 Thule
图勒

5 水 Water

Asteroids & Moons, Comets 99

Greek and Roman name of remote island in the North Atlantic 图勒是在古希腊和罗马时期北大西洋中的极北之地

---

100 Asteroids and Moons, Comets 小行星及其卫星和彗星

1583 Antilochus
安提洛克斯

4 金 Metal

Asteroids & Moons, Comets 100

Greek warrior killed in the Trojan War; son of Nestor
安提洛克斯是希腊将领，死于特洛伊战争，内斯特之子

101 Asteroids and Moons, Comets 小行星及其卫星和彗星

884 Priamus 普里阿摩斯星 **3** 土 Earth

Asteroids & Moons, Comets 101

King Priam of Troy pays a ransom to Achilles for his son's body 特洛伊国王普里阿摩斯向阿喀琉斯赎回儿子尸体

102 Asteroids and Moons, Comets 小行星及其卫星和彗星

1437 Diomedes 狄奥梅迪斯 **4** 金 Metal

Asteroids & Moons, Comets 102

Greek hero Diomedes steals the Palladium from Troy 希腊英雄狄奥梅迪斯从特洛伊城偷走了守护神雕像

103 Asteroids and Moons, Comets 小行星及其卫星和彗星

1749 Telamon 忒拉蒙 **4** 金 Metal

Asteroids & Moons, Comets 103

Greek king who helped Heracles capture Troy 希腊国王忒拉蒙帮助赫拉克勒斯夺取特洛伊城

104 Asteroids and Moons, Comets 小行星及其卫星和彗星

1172 Aeneas 埃涅阿斯 **5** 水 Water

Asteroids & Moons, Comets 104

The Tiber River God speaks to Aeneas in a dream 埃涅阿斯梦见与台伯河河神说话

105 Asteroids and Moons, Comets 小行星及其卫星和彗星	106 Asteroids and Moons, Comets 小行星及其卫星和彗星
588 **Achilles** 阿喀琉斯 4 金 Metal  Bravest Greek warrior in the Trojan War 阿喀琉斯是特洛伊战争中希腊第一勇士	659 **Nestor** 内斯特 4 金 Metal  Oldest Greek warrior in the Trojan War 特洛伊战争中最老的希腊勇士
107 Asteroids and Moons, Comets 小行星及其卫星和彗星	108 Asteroids and Moons, Comets 小行星及其卫星和彗星
911 **Agamemnon** 阿伽门农 4 金 Metal 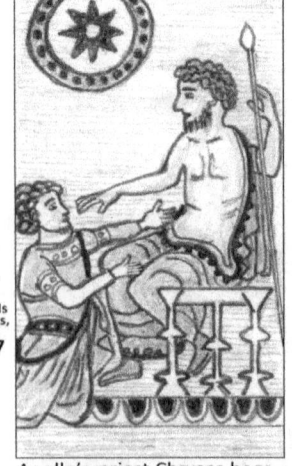 Apollo's priest Chryses begs Agamemnon to release his daughter Chryseis 阿波罗的祭司克律塞斯恳求阿伽门农释放他的女儿克律塞伊斯	624 **Hektor** 赫克特 4 金 Metal  Commander of the Trojan Army; son of King Priam 特洛伊战争中特洛伊城之统帅，普里阿摩斯的儿子

108/1 Asteroids and Moons, Comets 小行星及其卫星和彗星 S/2006(624)1	**109** Asteroids and Moons, Comets 小行星及其卫星和彗星 617 Patroclus 帕特罗克洛斯 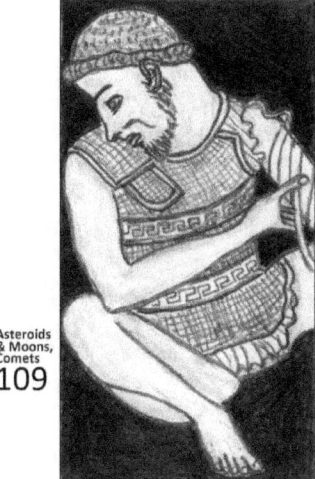 Asteroids & Moons, Comets 109 Greek warrior Patroclus wounded in the Trojan War 希腊勇士帕特罗克洛斯在特洛伊战争中受伤
109/1 Asteroids and Moons, Comets 小行星及其卫星和彗星 Menoetius 墨诺提俄斯　5水 Water  Asteroids & Moons, Comets 109/1 Father of Patroclus; one of the Argonauts 帕特罗克洛斯的父亲，阿尔戈英雄之一	**110** Asteroids and Moons, Comets 小行星及其卫星和彗星 1647 Menelaus 墨涅拉俄斯　4金 Metal  Asteroids & Moons, Comets 110 King of Sparta and husband of Helen 斯马达国王，美女海伦的丈夫

111 Asteroids and Moons, Comets 小行星及其卫星和彗星

1208 Troilus 特洛伊罗斯

Constellation **80**

Planets & Moons
6/39, 7/20, 9/3

Asteroids & Moons, Comets **111**

Lunar Mansion **25**

Earthly Branch **7** 午马 Horse

Son of King Priam of Troy; murdered by Achilles
普里阿摩斯之子，被阿喀琉斯所杀

---

112 Asteroids and Moons, Comets 小行星及其卫星和彗星

1143 Odysseus 奥德修斯 **5** 水 Water

Asteroids & Moons, Comets **112**

Odysseus
(Ulysses 尤利西斯)

Greek sailor, adventurer and Trojan War veteran 希腊冒险家，特洛伊战争勇士

---

113 Asteroids and Moons, Comets 小行星及其卫星和彗星

1404 Ajax 大埃阿斯星 **4** 金 Metal

Asteroids & Moons, Comets **113**

Greek warrior in the Trojan War; son of Telamon 特洛伊战争勇士，忒拉蒙之子

---

114 Asteroids and Moons, Comets 小行星及其卫星和彗星

1173 Anchises 安喀塞斯 **2** 火 Fire

Asteroids & Moons, Comets **114**

Trojan warrior Aeneas carries his crippled father Anchises
特洛伊战争勇士埃涅阿斯背着他的残疾父亲安喀塞斯

115 Asteroids and Moons, Comets 小行星及其卫星和彗星 **944 Hidalgo** 希达尔戈  Asteroids & Moons, Comets **115** Miguel Hidalgo y Costilla Mexican Revolutionary Leader 墨西哥革命领袖米格尔·希达尔戈·伊·科斯蒂利亚	**116** Asteroids and Moons, Comets 小行星及其卫星和彗星 **27P/ Crommelin (Comet)** 克罗玛林彗星  Asteroids & Moons, Comets **116** Andrew Claude de la Cherois Crommelin (1865-1939) 爱尔兰天文学家安德鲁·克罗姆林
117 Asteroids and Moons, Comets 小行星及其卫星和彗星 **2060 Chiron** 凯龙星 Conste-llation 49, 70 **95P/Chiron** 凯龙彗星 (Comet)  Asteroids & Moons, Comets **117** Son of Cronos; wisest of the centaurs 克洛诺斯的儿子，马人中凯龙是最有智能者	**118** Asteroids and Moons, Comets 小行星及其卫星和彗星 **10199 Chariklo** 女凯龙星  Asteroids & Moons, Comets **118** Wife of the centaur Chiron 半人马凯龙(喀戎)的妻子

119	Asteroids and Moons, Comets 小行星及其卫星和彗星

1P/ Halley (Comet)
哈雷彗星

Asteroids & Moons, Comets
119

English Astronomer Edmond Halley (1656-1742)
英国天文学家爱德蒙·哈雷

120	Asteroids and Moons, Comets 小行星及其卫星和彗星

5145 Pholus
人龙星

Conste-llation 49, 70

Asteroids & Moons, Comets
120

The centaur Pholus holds a jar of wine
马人福洛斯拿着一罐酒

121	Asteroids and Moons, Comets 小行星及其卫星和彗星

7066 Nessus
毒龙星 Water

Conste-llation 49, 70

Asteroids & Moons, Comets
121

Nessus carries Hercules' wife Deianira across the river
马人涅嗦斯驮着赫拉克勒斯的妻子得伊阿尼拉过河

122	Asteroids and Moons, Comets 小行星及其卫星和彗星

C1907/G1 Grigg-Mellish
格里格-墨丽丝彗星 (Comet)

J.E. Mellish

Asteroids & Moons, Comets
122

John Edward Mellish (1886-1970)
约翰·爱德华·墨丽丝
John Geigg (1838-1920)
约翰·格里格

123 Asteroids and Moons, Comets 小行星及其卫星和彗星 **42355** Typhon 提丰  Dangerous monster who attacked the gods 恐怖的怪兽提丰袭击主神宙斯	**123/1** Asteroids and Moons, Comets 小行星及其卫星和彗星 Echidna 厄客德娜  Wife of Typhon; mother of all monsters 提丰的妻子，所有魔兽的母亲
124 Asteroids and Moons, Comets 小行星及其卫星和彗星 **65489** Ceto 刻托 5水 Water Constellation 6 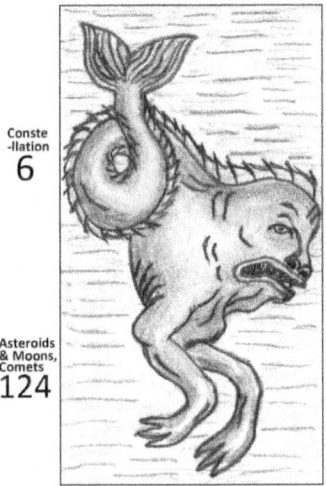 Greek Sea Goddess; mother of sea monsters 希腊神话中海神，海怪的母亲，福耳库斯的妻子	**124/1** Asteroids and Moons, Comets 小行星及其卫星和彗星 Phorcys 福耳库斯 5水 Water  Greek Sea God; husband of Ceto; father of monsters 希腊神话中原始海神，刻托丈夫，海怪的父亲

SUIT 1, Wood 木 (Air 空气) = Swords 剑, Spades 黑桃, Tens 十 [10s]

The Tarot suit of Swords/Spades is equivalent to the 1st Chinese element Wood (or Greek element Air). Cards 1 to 9 are valued in the Tens and on them are depicted warriors and swordswomen from the novel Shuǐ Hǔ Zhuàn 水浒传 (the Water Margin). Card 10 represents the Phoenix Fèng 凤 found in mahjong sets in the first decade of the 20th century before it was replaced by the green 发 Fā. China's first female military commander Fù Hǎo 妇好 who died c. 1200 BC shares Card 10 with the phoenix. The suit of Swords reintroduces the suit of Tens (Shízì 十字) which was dropped from Mǎ Diào 马吊 paper cards during the reign (1736-1795) of Emperor Qián Lóng 乾隆. The Chinese character 十 Shí (Ten) looks a little like a sword with the hilt above and the blade below and it may have inspired the Italian and Tarot suit of Swords.

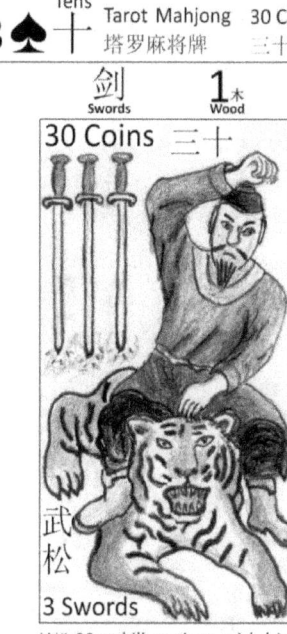

Wǔ Sōng kills a tiger with his bare hands

Ruǎn Xiǎo Èr

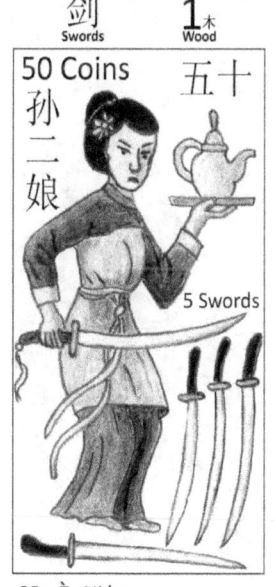

Sūn Èr Niáng

Yàn Qīng

11 ♠ Tarot Mahjong 塔罗麻将牌

剑 Swords　1木 Wood

Bodyguard of King　侍卫

12 ♠ Tarot Mahjong 塔罗麻将牌

剑 Swords　1木 Wood

Knight　骑士

13 ♠ Tarot Mahjong 塔罗麻将牌

剑 Swords　1木 Wood

Queen　女王

卞氏

魏女王

Biàn Shì Queen of Wèi

14 ♠ Tarot Mahjong 塔罗麻将牌

剑 Swords　1木 Wood

King　国王

曹操

魏王

Cáo Cāo (155-220) King of Wèi

SUIT 2, Fire 火 = Batons 棍, Clubs 草花, Sticks 条 (Bamboo 竹), Hundreds 百 [100s]

The Tarot suit of Batons 棍, or Clubs 草花 is equivalent to the 2nd Chinese Element Fire, the ancient Greek element Fire and the Mahjong suit of Tiáo 条 (Sticks) or Suǒ 索 (Rope). Suǒ 索 is the rope passed through the square holes in ancient coins to make a string of coins but because it is straight and long, it is also called Tiáo 条 (條) meaning a Stick and to English mahjong players it is the suit of Bamboo. Card 1 displays the mahjong Yāo Jī bird (幺鸡). Card 10 represents the mahjong tile 发 Fā or 发财 Fā Cái "make a fortune". This card shows 10 strings of 100 coins making a total of 1,000 coins and an early 20th century female bank teller using an abacus. It replaces Lǎo Qiān 老千 (Old Thousand) of the 18th Century Pèng Hé paper cards 碰和牌.

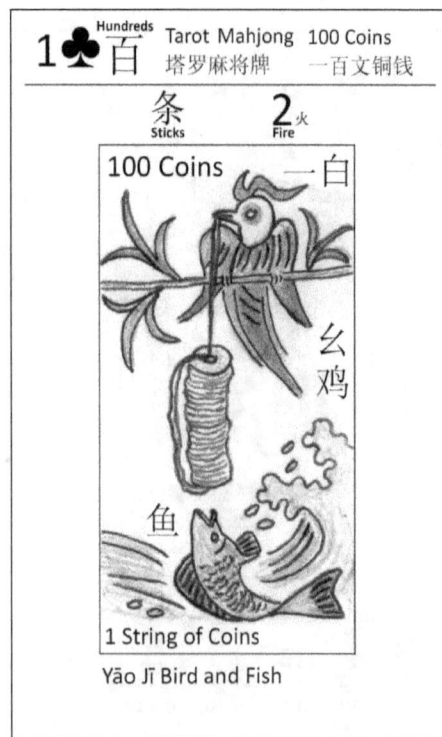

Yāo Jī Bird and Fish

Lǎo Zǐ (6th Century B.C.) Daoist Philosopher

3♣ 百	Tarot Mahjong 300 Coins 塔罗麻将牌 三百文铜钱

条 Sticks   2 火 Fire

300 Coins 三百 庄子 3 Sticks 三条

Zhuāng Zǐ (c. 370-301 B.C.)
Daoist Philosopher

4♣ 百	Tarot Mahjong 400 Coins 塔罗麻将牌 四百文铜钱

条 Sticks   2 火 Fire

400 Coins 四百 墨子 4 Sticks 四条

Mò Zǐ (470-391 B.C.)
Universal Love Philosopher

5♣ 百	Tarot Mahjong 500 Coins 塔罗麻将牌 五百文铜钱

条 Sticks   2 火 Fire

500 Coins 五百 孔子 5 Sticks 五条

Kǒng Zǐ (Confucius)
551-479 B.C.

6♣ 百	Tarot Mahjong 600 Coins 塔罗麻将牌 六百文铜钱

条 Sticks   2 火 Fire

600 Coins 六百 孟子 6 Sticks 六百

Mèng Zǐ (Mencius) 327-289 B.C.
Confucian Philosopher

7♣ 百 Tarot Mahjong 700 Coins
塔罗麻将牌　七百文铜钱

8 Sticks　条 Sticks　2 火 Fire

700 Coins　七百
7 Sticks　七条

Xún Zǐ (c. 312-230 B.C.)
Confucian Philosopher

8♣ 百 Tarot Mahjong 800 Coins
塔罗麻将牌　八百文铜钱

8 Sticks　条 Sticks　2 火 Fire

800 Coins　八百
8 Sticks　八条

Hán Fēi Zǐ (c. 280-233 B.C.)
Legalist Philosopher

9♣ 百 Tarot Mahjong 900 Coins
塔罗麻将牌　九百文铜钱

条 Sticks　2 火 Fire

900 Coins　九百
熊猫
竹子
9 Sticks　九条

Panda and Bamboo

10♣ 百 Hundreds Tarot Mahjong 塔罗麻将牌
Centre, Wealth, White, Dragon, Phoenix
中发白龙凤

发 Wealth　条 Sticks　2 火 Fire

壹贯
1,000 Coins

10 Strings of Coins
Making a Fortune　发财

11 ♣ Tarot Mahjong 塔罗麻将牌

条 Sticks　2 火 Fire

Bodyguard of King　侍卫

12 ♣ Tarot Mahjong 塔罗麻将牌

条 Sticks　2 火 Fire

Knight　骑士

13 ♣ Tarot Mahjong 塔罗麻将牌

条 Sticks　2 火 Fire

Queen　女王

甘夫人

蜀女王

Gān Fū Ren, Queen of Shǔ

14 ♣ Tarot Mahjong 塔罗麻将牌

条 Sticks　2 火 Fire

King　国王

刘备

蜀王

Liú Bèi (161-223)
King of Shǔ

SUIT 3, Earth 土 = Coins 筒 (硬币), Diamonds 方片, Single Coins [Units]

The Tarot suit of Coins/Diamonds is equivalent to the 3rd Chinese element Earth, the ancient Greek element Earth and the Mahjong suit of 筒 Tǒng (Coins, cylinders or Circles, sometimes called Bǐng 饼 which is a round moon cake). The suit of Coins or Circles 圆 in Italian playing cards and in Tarot cards is an exact equivalent of this Mahjong suit. Cards 1 to 9 have the value of single coins or units. Card 10 represents the mahjong tile 白 Bái (White) but in this tarot pack the female outlaw fighter 白花 Bái Huā (White Flower) otherwise known as 一丈青 Yī Zhàng Qīng or 扈三娘 Hù Sān Niáng is placed inside the framed white mahjong rectangle. This continues the tradition of the honor card White Flower in the 18th Century Pèng Hé paper cards 碰和牌.

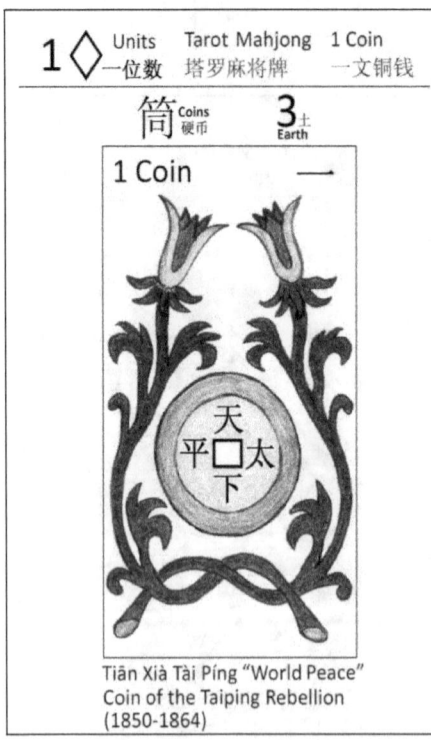

Tiān Xià Tài Píng "World Peace" Coin of the Taiping Rebellion (1850-1864)

Dragon and Phoenix coinage of Emperor Yuán Huì Zōng 元惠宗 (Toghun Temür) 1320-1370

3 ◊ Units 一位数	Tarot Mahjong 塔罗麻将牌	3 Coins 三文铜钱	4 ◊ Units 一位数	Tarot Mahjong 塔罗麻将牌	4 Coins 四文铜钱

筒 Coins 硬币 3 土 Earth

Grand Canal 1776 km long completed in 6th Century

筒 Coins 硬币 3 土 Earth

Merchants on the Silk Road

5 ◊ Units 一位数	Tarot Mahjong 塔罗麻将牌	5 Coins 五文铜钱	6 ◊ Units 一位数	Tarot Mahjong 塔罗麻将牌	6 Coins 六文铜钱

筒 Coins 硬币 3 土 Earth

Western end of the Great Wall and Silk Road at Jiā Yù Guān

筒 Coins 硬币 3 土 Earth

Fǔ Jìn Mén Gate in Shěnyáng City

7 ◊	Units 一位数	Tarot Mahjong 塔罗麻将牌	7 Coins 七文铜钱

筒 Coins 硬币 3 土 Earth

7 Coins 七 白塔

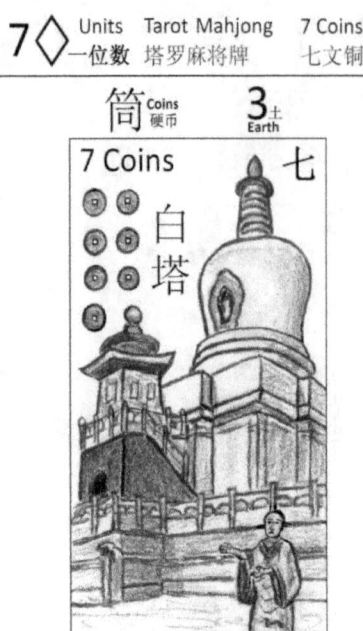

The White Pagoda, Běihǎi Park, Běijīng

8 ◊	Units 一位数	Tarot Mahjong 塔罗麻将牌	8 Coins 八文铜钱

筒 Coins 硬币 3 土 Earth

8 Coins 八 摇钱树 土地神

Money Tree and the Immortal of the Earth

9 ◊	Units 一位数	Tarot Mahjong 塔罗麻将牌	9 Coins 九文铜钱

9 Coins 九 天坛

Temple of Heaven, Běijīng

10 ◊	Units 一位数	Tarot Mahjong 塔罗麻将牌 Centre, Wealth, White, Dragon, Phoenix 中发白龙凤

白 White 筒 Coins 硬币 3 土 Earth

10 Coins

White Flower
Yī Zhàng Qīng

Fán Jī, Queen of Chǔ

Chǔ Zhuāng Wáng, King of Chǔ.
Reigned 613-591 B.C.

SUIT 4, Metal 金 (Aether/Spirit 苍天/精神世界) = The Major Arcana 大秘仪, Towers, Thousands 千 [1,000s]

The Tarot Major Arcana suit (大秘仪 Dà-mì-yí or 大啊尔克那 Dà Ā-ěr-kè-nà) is represented by watch towers on the Great Wall 瞭望塔 Liào Wàng Tǎ or guard towers 长城岗楼 Cháng Chéng Gǎng Lóu and is equivalent to the 4th Chinese element Metal and the ancient Greek element Aether/Spirit which is the non-material world or spirit world. Cards 1 to 9 are valued in thousands. Card 22 represents the character 龙 Lóng (Dragon) which appeared in some mahjong sets in the first decade of the 20th century before it was replaced by the red character 中 Zhōng.

2 🏯 千 Tarot Mahjong 2,000 Coins 塔罗麻将牌 二千文铜钱 大秘仪 Major Arcana 4 金 Metal  Golden Mother of Lake Yáo in the Kūn Lún Mountains	**3** 🏯 千 Tarot Mahjong 3,000 Coins 塔罗麻将牌 三千文铜钱 大秘仪 Major Arcana 4 金 Metal  Wǔ Zé Tiān (624-705) Táng Dynasty
4 🏯 千 Tarot Mahjong 4,000 Coins 塔罗麻将牌 四千文铜钱 大秘仪 Major Arcana 4 金 Metal  Táng Gāo Zōng (628-683)	**5** 🏯 千 Tarot Mahjong 5,000 Coins 塔罗麻将牌 五千文铜钱 大秘仪 Major Arcana 4 金 Metal  Zhāng Dào Líng (34-156) Daoist Celestial Master

6 Tarot Mahjong 6,000 Coins
塔罗麻将牌 六千文铜钱

大秘仪 Major Arcana 4 金 Metal

The Lovers 情人
织女
牛郎

Gemini 双子座

Zhī Nǚ meets Niú Láng on the Magpie Bridge

7 Tarot Mahjong 7,000 Coins
塔罗麻将牌 七千文铜钱

大秘仪 Major Arcana 4 金 Metal

The Chariot 战车

Cancer 巨蟹座

Han Dynasty War Chariot

8 Tarot Mahjong 8,000 Coins
塔罗麻将牌 八千文铜钱

大秘仪 Major Arcana 4 金 Metal

Strength 力量
中国银行

Leo 狮子座

Bank of China, founded in 1912. Financial Strength

9 Tarot Mahjong 9,000 Coins
塔罗麻将牌 九千文铜钱

大秘仪 Major Arcana 4 金 Metal

The Hermit 隐士
诗人林逋

Virgo 室女座

Lín Bū (967-1028) roams the mountains in moonlight

10 长城 Great Wall Guard Tower Tarot Mahjong 塔罗麻将牌 大秘仪 Major Arcana 4 金 Metal Wheel of Life 大轮回 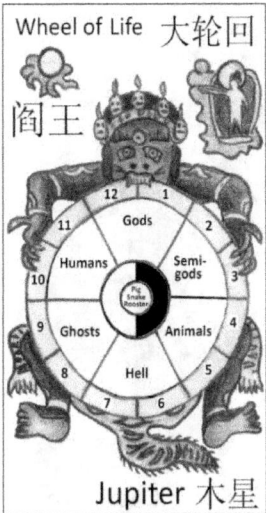 Jupiter 木星 Yán Wáng (Yama) turns the Tibetan Wheel of Life	**11** 长城 Great Wall Guard Tower Tarot Mahjong 塔罗麻将牌 大秘仪 Major Arcana 4 金 Metal Justice 正义 天平 Bì Àn 犴 Dragon Dog 獬豸 Libra 天秤座 Bì Àn, Xiè Zhì (Unicorn) and Balance — Symbols of Justice
12 长城 Great Wall Guard Tower Tarot Mahjong 塔罗麻将牌 大秘仪 Major Arcana 4 金 Metal The Hanged Man 吊死的人  Water 水 Last Ming Emperor Chóng Zhēn (1611-1644) hangs himself	**13** 长城 Great Wall Guard Tower Tarot Mahjong 塔罗麻将牌 大秘仪 Major Arcana 4 金 Metal Death 死亡 67,000 dead in Battle of Mukden 1905  Scorpio 天蝎座 Foreign invaders kill each other in China in 1905

14 Tarot Mahjong 塔罗麻将牌	**15** Tarot Mahjong 塔罗麻将牌
大秘仪 Major Arcana — 4 金 Metal	大秘仪 Major Arcana — 4 金 Metal

Temperance 节制 — Sagittarius 人马座 Scholars drink moderately on a mountain	The Devil 恶魔 魔障 — Capricorn 摩羯座 The evil Mó Zhàng (Mara) attacks the Buddha
16 Tarot Mahjong 塔罗麻将牌	**17** Tarot Mahjong 塔罗麻将牌
大秘仪 Major Arcana — 4 金 Metal	大秘仪 Major Arcana — 4 金 Metal

The Tower 雷峰塔 西湖 — Mars 火星 Léi Fēng Pagoda near Hángzhōu collapsed in 1924	The Star 星 斗母元君 北斗九真圣德天后 — Aquarius 水瓶座 Dǒu Mǔ, Goddess of the North Star, Queen of Heaven

18 长城阃楼 Great Wall Guard Tower Tarot Mahjong 塔罗麻将牌 大秘仪 Major Arcana 4 金 Metal  Cháng É ascends to the moon	**19** 长城阃楼 Great Wall Guard Tower Tarot Mahjong 塔罗麻将牌 大秘仪 Major Arcana 4 金 Metal  Hòu Yì shoots down 9 of the 10 suns and their sun-birds
20 长城阃楼 Great Wall Guard Tower Tarot Mahjong 塔罗麻将牌 大秘仪 Major Arcana 4 金 Metal  Bāo Zhěng (999-1062) judges affairs in the afterlife	**21** 长城阃楼 Great Wall Guard Tower Tarot Mahjong 塔罗麻将牌 大秘仪 Major Arcana 4 金 Metal 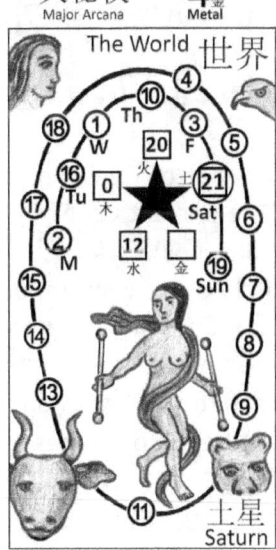 12 Signs of the Zodiac, 7 Days and 5 Elements

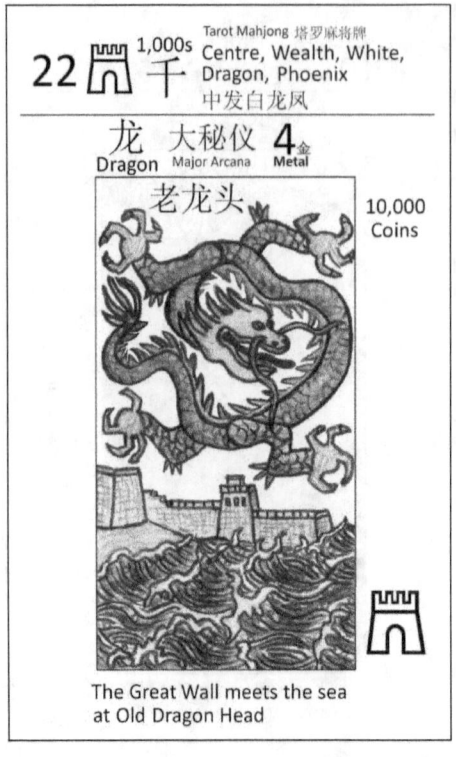

Suit 5, Water 水 = Cups 杯子, *Wàn* 万, Hearts 红桃, Ten Thousands [10,000s]

The Tarot suit of Cups/Hearts is equivalent to the 5th Chinese element Water, the ancient Greek element Water and the Mahjong suit of *Wàn* (Ten Thousands). Wàn (万 or 萬) looks like a cup upside-down and Marco Polo or his contemporaries may have misinterpreted it and given us a suit of Cups in Tarot and Italian cards. Cards 1 to 9 are valued in the tens of thousands and mostly depict ancient Chinese mariners. Card 10 represents the red mahjong character Zhōng 中 (Centre) as in Zhōng Guó 中国 (China or Chinese) and it shows China's Ancient Maritime Silk Road 中国古代海上丝绸之路. When mahjong was invented in the mid to late 19th Century the red Zhōng 中 replaced the honor card Red Flower (红花 Hóng Huā) of the 18th Century Pèng Hé cards 碰和牌.

Black Tiger Spring, Jínán City, Shāndōng Province

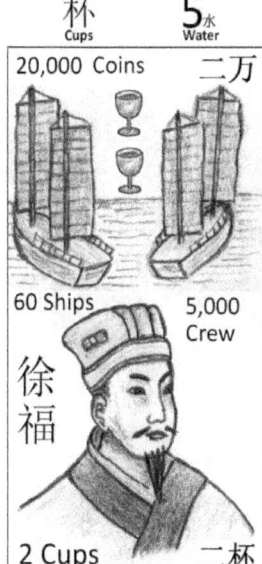

Xú Fú (born 255 B.C.) sails Pacific Ocean in 210 B.C.

Fǎ Xiǎn (337-c. 422 A.D.) returns to China by ship

Huì Shēn returns from Pacific Ocean voyage in 499 A.D.

5♡万 Tarot Mahjong 50,000 Coins 塔罗麻将牌 五万文铜钱  Xuán Zàng (602-664) travels to India	**6♡万** Tarot Mahjong 60,000 Coins 塔罗麻将牌 六万文铜钱  Jiàn Zhēn (688-763) sails to Japan
7♡万 Tarot Mahjong 70,000 Coins 塔罗麻将牌 七万文铜钱  Wāng Dà Yuān (fl. 1311-1350) sails to India and Africa	**8♡万** Tarot Mahjong 80,000 Coins 塔罗麻将牌 八万文铜钱  Zhèng Hé (1371-1433) sails to India, Arabia and Africa

9♥万 Tarot Mahjong 90,000 Coins
塔罗麻将牌 九万文铜钱
杯 Cups 5水 Water

90,000 Coins 九万
长颈鹿
9 Cups 九杯

Giraffe brought to China from Somalia in 1415

10♥万 10,000s Tarot Mahjong 塔罗麻将牌
Centre, Wealth, White, Dragon, Phoenix
中发白龙凤

中 Centre 杯 Cups 5水 Water

100,000 Coins
中

Chinese Ancient Maritime Silk Road 中国古代海上丝绸之路

11♥ Tarot Mahjong 塔罗麻将牌
杯 Cups 5水 Water

Bodyguard of King 侍卫

12♥ Tarot Mahjong 塔罗麻将牌
杯 Cups 5水 Water

Knight 骑士

The Four Great Heavenly Kings 四大天王 and the Earth God 土地神 Tǔ Dì Shén

The Four Kings represent the 4 mahjong directional tiles 东 Dōng, 南 Nán, 西 Xī, 北 Běi (East, South, West, North) known in English as the 4 Winds. However, because there are 5 suits we need the Earth God Tǔ Dì Shén to represent Suit 3 Earth (Coins/Diamonds) which relates to the Centre. Images of the Four Great Heavenly Kings can be seen in most Chinese temples where they also represent the four directions.

Tarot Mahjong 塔罗麻将牌	Tarot Mahjong 塔罗麻将牌
15 ♠ Earth God and 4 Great Heavenly Kings 土地神和四大天王	**15** ♣ Earth God and 4 Great Heavenly Kings 土地神和四大天王

1 木 Wood — 东方 East

持国天王 调

Heavenly King of the East:
Well suited

2 火 Fire — 南方 South

增长天王 风

Heavenly King of the South:
Wind

Tarot Mahjong 塔罗麻将牌	Tarot Mahjong 塔罗麻将牌
15 ♦ Earth God and 4 Great Heavenly Kings 土地神和四大天王	**23** 长城守护塔 Great Wall Guard Tower Earth God and 4 Great Heavenly Kings 土地神和四大天王

3 土 Earth — 中央 Centre

Earth God Tǔ Dì Shén

土地神 土

4 金 Metal — 西方 West

广目天王 顺

Heavenly King of the West:
Smooth

The Five Great Ancient Capitals of China 中国五大古都: Běijīng 北京, Nánjīng 南京, Luòyáng 洛阳, Cháng'ān 长安 (now known as Xī'ān 西安) and Kāifēng 开封

Each of the 5 capitals is in Suit 3 Earth (Coins/Diamonds) because the 5 mahjong groups (the 4 Seasons, the 4 Flowers, the 4 Pastimes, the 4 Professions and the 4 Animals) lack the 3rd Element Earth which represents the Change of Seasons and the Centre. In ancient China the Earth God usually resided near the city gate.

16 ♦ Tarot Mahjong 塔罗麻将牌 **Five Great Ancient Capitals** 中国五大古都	17 ♦ Tarot Mahjong 塔罗麻将牌 **Five Great Ancient Capitals** 中国五大古都
3 Earth 北京 Běijīng  正阳门 前门 Zhèng Yáng Mén Gate now known as Qián Mén	3 Earth 南京 Nánjīng  中华门 Zhōng Huá Mén The China Gate

18 ♦ Tarot Mahjong 塔罗麻将牌 **Five Great Ancient Capitals** 中国五大古都	19 ♦ Tarot Mahjong 塔罗麻将牌 **Five Great Ancient Capitals** 中国五大古都
3 Earth 洛阳 Luòyáng  丽景门 Lì Jǐng Gate	3 Earth 长安 Cháng'ān 西安北门  安远门 Ān Yuǎn Gate in Xī'ān City previously called Cháng'ān

The Four Seasons, Four Flowers, Four Pastimes, Four Occupations and Four Animals from Mahjong

The Four Seasons: Spring, Summer, Autumn, Winter (春夏秋冬 Chūn Xià Qiū Dōng); the Four Flowers: Plum, Orchid, Bamboo, Chrysanthemum (梅 兰 竹 菊 Méi Lán Zhú Jú); the Four Pastimes: Music, Chess, Calligraphy or Literacy, Painting (琴棋书画 Qín Qí Shū Huà); the Four Professions: Fisherman, Woodcutter, Farmer, Scholar (鱼樵耕读 Yú Qiáo Gēng Dú); the 4 Mahjong Animals: Mouse, Cat, Centipede, Cockerel (老鼠 Lǎo Shǔ, 猫 Māo, 蜈蚣 Wú Gōng, 鸡 Jī). Each of these 5 groups lacks the 3rd Element Earth which represents the Change of Seasons and the Centre. Therefore the Five Great Ancient Capitals of China (中国五大古都 Zhōng Guó Wǔ Dà Gǔ Dū) make up for these 5 missing cards in the 3rd suit Earth (Coins/Diamonds).

16 ♠	Tarot Mahjong 塔罗麻将牌 Four Seasons 四季	16 ♣	Tarot Mahjong 塔罗麻将牌 Four Seasons 四季

1 木 Wood

春桃 Spring

Peach Blossom

2 火 Fire

夏荷 Summer

Lotus

24  Great Wall Guard Tower	Tarot Mahjong 塔罗麻将牌 Four Seasons 四季	16 ♥	Tarot Mahjong 塔罗麻将牌 Four Seasons 四季

4 金 Metal

秋菊 Autumn

Chrysanthemum

5 水 Water

冬梅 Winter

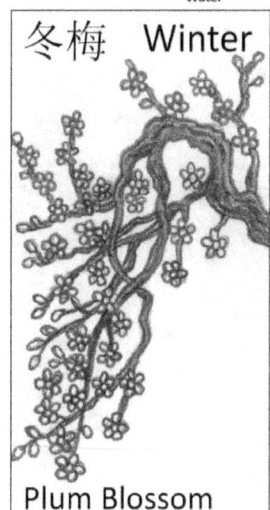

Plum Blossom

17 ♠ Tarot Mahjong 塔罗麻将牌
Four Flowers 四君子花

1 木 Wood

兰 Orchid

Spring 春

17 ♣ Tarot Mahjong 塔罗麻将牌
Four Flowers 四君子花

2 火 Fire

竹 Bamboo

Summer 夏

25  Great Wall Guard Tower — Tarot Mahjong 塔罗麻将牌
Four Flowers 四君子花

4 金 Metal

菊 Chrysanthemum

Autumn 秋

17 ♥ Tarot Mahjong 塔罗麻将牌
Four Flowers 四君子花

5 水 Water

梅 Plum Blossom

Winter 冬

18 ♠	Tarot Mahjong 塔罗麻将牌 Four Pastimes 琴棋书画	18 ♣	Tarot Mahjong 塔罗麻将牌 Four Pastimes 琴棋书画
1 木 Wood	琴 Music	2 火 Fire	棋 Chess

26  Great Wall Guard Tower	Tarot Mahjong 塔罗麻将牌 Four Pastimes 琴棋书画	18	Tarot Mahjong 塔罗麻将牌 Four Pastimes 琴棋书画
4 金 Metal	书 Calligraphy	5 水 Water	画 Painting

19 ♠	Tarot Mahjong 塔罗麻将牌 Four Professions 渔樵耕读	19 ♣	Tarot Mahjong 塔罗麻将牌 Four Professions 渔樵耕读

1 木 Wood — 渔 Fisherman

2 火 Fire — 樵 Woodcutter

27 Great Wall Guard Tower	Tarot Mahjong 塔罗麻将牌 Four Professions 渔樵耕读	19 ♥	Tarot Mahjong 塔罗麻将牌 Four Professions 渔樵耕读

4 金 Metal — 耕 Farmer

5 水 Water — 读 Scholar

20 ♠	Tarot Mahjong 塔罗麻将牌 Four Mahjong Animals 四麻将动物

1 木 Wood

猫 Cat

20 	Tarot Mahjong 塔罗麻将牌 Four Mahjong Animals 四麻将动物

2 火 Fire

蜈蚣 Centipede

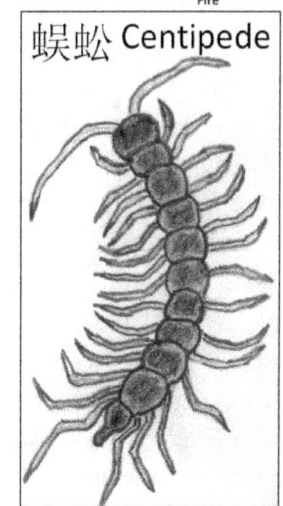

28	Tarot Mahjong 塔罗麻将牌 Four Mahjong Animals 四麻将动物

4 金 Metal

鸡 Cockerel

20	Tarot Mahjong 塔罗麻将牌 Four Mahjong Animals 四麻将动物

5 水 Water

老鼠 Mouse

The 5 Stars of Good Fortune 五星

Fú Xīng 福星 the Star of Blessings and Good Fortune wears the winged hat of a court official and carries a curved jade wishing bar known as a Rú Yì 如意. He is usually accompanied by a bat (Biān Fú 蝙蝠), the second character Fú being a homophone of his name. Lù Xīng 禄星, the star of prosperity and official salary, holds the Wealth-attracting Child (Zhāo Cái Tóng Zǐ 招财童子) and is accompanied by a deer (Lù 鹿) which is a homophone of his name. Shòu Xīng 寿星, the Star of Longevity, has an enormous head and holds the Peach of Immortality (Shòu Táo 寿桃). He carries a dragon-headed peach-wood staff and is accompanied by a crane (Hè 鹤) which symbolizes longevity. Xǐ Shén 喜神 is the God of Happiness and holds a type of golden boat-shaped money or golden treasure ingot. He is accompanied by a Magpie (Xǐ Què 喜鹊), the first character being a homophone of his name. Cái Shén 财神 is the God of Wealth and his treasure bowl contains, among other things, a fish Yú 鱼, a homophone of Yú 余 meaning "surplus".

Tarot Mahjong 塔罗麻将牌	
21	**Five Stars of Good Fortune** 五星

3. 寿星　Shòu Xīng　　3 土 Earth

Star of Longevity holds the Peach of Immortality

Tarot Mahjong 塔罗麻将牌	
29	**Five Stars of Good Fortune** 五星

4. 喜神　Xǐ Shén　　4 金 Metal

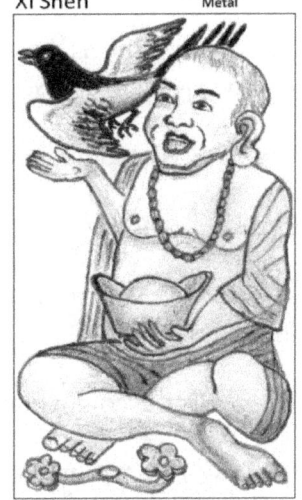

God of Happiness holds a Golden Treasure Ingot

Tarot Mahjong 塔罗麻将牌	
21	**Five Stars of Good Fortune** 五星

5. 财神　Cái Shén　　5 水 Water

The God of Wealth displays his Treasure Bowl

The 5 Animals of the Fortune Stars 五星对应的动物

Three of these animals have names which are homophones of the names of the Five Fortune Stars Fú, Lù Shòu, Xǐ, Cái.

1. Fú Xīng 福星 is accompanied by the bat Biān Fú 蝙蝠.

2. Lù Xīng 禄星 is accompanied by the deer Lù 鹿.

3. Shòu Xīng 寿星 is accompanied by the crane Hè 鹤 which symbolizes long life or immortality.

4. Xǐ Shén 喜神 is accompanied by the magpie Xǐ Què 喜鹊.

5. Cái Shén 财神 has, among other things in his Treasure Bowl, a fish Yú 鱼, a homophone of the God of Wealth's attribute Yú 余 "surplus".

| 22 | Tarot Mahjong 塔罗麻将牌 **Animals of the Five Fortune Stars** 五星对应的动物 |

3. 鹤
He (Crane)   **3** 上 Earth

Hè 鹤 (Crane) represents Shòu Xīng 寿星 the Star of Longevity

| 30 | Tarot Mahjong 塔罗麻将牌 **Animals of the Five Fortune Stars** 五星对应的动物 |

Great Wall Guard Tower 长城守城楼

4. 喜鹊
Xǐ Què (Magpie)   **4** 金 Metal

Xǐ Què 喜鹊 (Magpie) represents Xǐ Shén 喜神 the God of Happiness

| 22 | Tarot Mahjong 塔罗麻将牌 **Animals of the Five Fortune Stars** 五星对应的动物 |

5. 鱼
Yú (Fish)   **5** 水 Water

Yú 鱼 (Fish) sounds like Yú 余 (Surplus) and represents the God of Wealth Cái Shén 财神

The 5 Symbols of Prosperity 五个吉祥物

The Five Symbols of Prosperity are intended to bring good fortune. 1. The curved jade wishing bar Rú Yì 如意 carried by Fú Xīng 福星 is an implement used for making your wishes come true. 2. The Wealth-attracting Child carried by Lù Xīng 禄星 can bring in money and take care of parents when they are old. 3. The Peaches of Immortality (Shòu Táo 寿桃) carried by Shòu Xīng 寿星 can bring long life. 4. The golden boat-shaped ingot carried by Xǐ Shén 喜神 signifies wealth, which can bring security and happiness. 5. The Treasure Bowl of Cái Shén ensures that farmers have a surplus of produce bringing security and happiness. It contains a fish Yú 鱼, a homophone of Yú 余 "surplus" as well as golden boat-shaped money ingots and other wealth related objects.

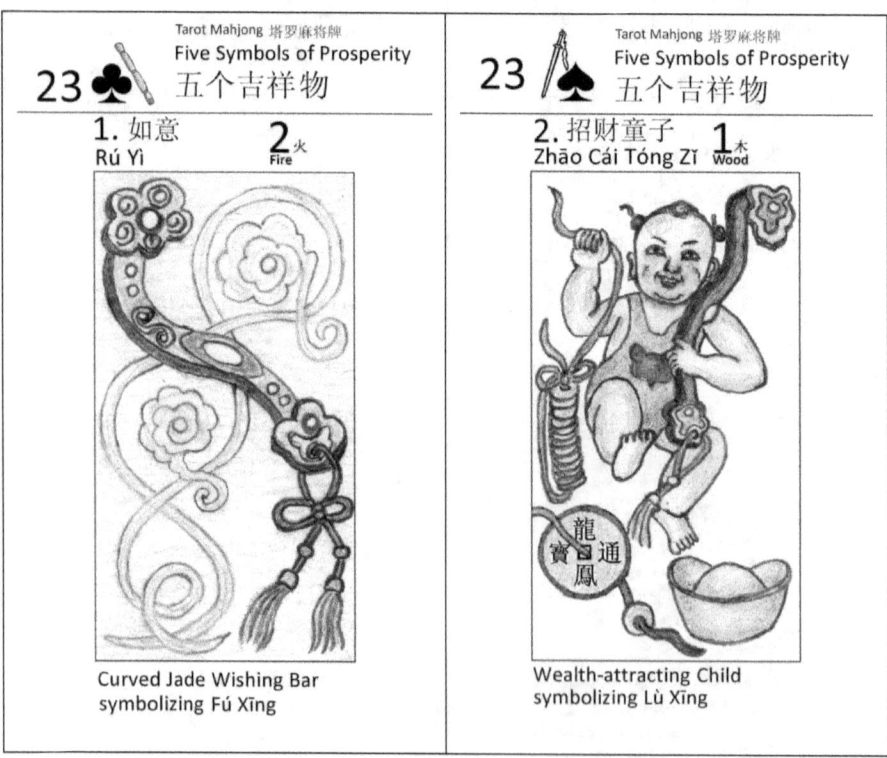

Tarot Mahjong 塔罗麻将牌

23 ◇ **Five Symbols of Prosperity**
五个吉祥物

3. 桃子
Táo Zi (Peaches)   **3** 土 Earth

Peaches of Immortality symbolizing Shòu Xīng

Tarot Mahjong 塔罗麻将牌

31 长城 Great Wall / 河楼 Guard Tower **Five Symbols of Prosperity**
五个吉祥物

4. 金元宝
Jīn Yuán Bǎo   **4** 金 Metal

Golden Treasure Ingots symbolizing Xǐ Shén.
Golden Boat-shaped Money.

Tarot Mahjong 塔罗麻将牌

23 ♡ **Five Symbols of Prosperity**
五个吉祥物

5. 聚宝盆
Jù Bǎo Pén   **5** 水 Water

Treasure Bowl symbolizing Cái Shén

The 5 Auspicious Animals 五瑞兽

The Fabulous animal Qí Lín 麒麟 brings prosperity and its appearance signifies that a nation has beneficent and just rulers. The other four animals bring in money, wealth and good fortune. The Money Toad Jīn Chán 金蟾 carries the Bā Guà 八卦 (Eight Trigrams) on its head and the seven stars of the Northern Dipper on its back. It spits out coins and strings of coins. The Wealth-attracting Cat Zhāo Cái Māo 招财猫 has a raised foreleg like the arm of a poker machine and produces a multitude of coins to enrich its devotees. The Pí Xiū 貔貅 swallows money and stores it safely in its body so that its devotees accumulate wealth. Some claim it is the 9th son of the dragon. Lóng Guī 龙龟 (Dragon Tortoise) is also a money- producing animal and possibly one of the sons of the dragon.

24 ♦ Tarot Mahjong 塔罗麻将牌 — Five Auspicious Animals 五瑞兽

3. 招财猫 Zhāo Cái Māo — 3 Earth

Fortune Cat with Gold Ingots and Coins

---

32 ⛩ Tarot Mahjong 塔罗麻将牌 — Five Auspicious Animals 五瑞兽
(Great Wall Guard Tower)

4. 貔貅 Pí Xiū — 4 Metal

Pí Xiū, also called Tiān Lù 天禄, has a voracious appetite for gold and silver and is a wealth-bringing creature.

---

24 ♡ Tarot Mahjong 塔罗麻将牌 — Five Auspicious Animals 五瑞兽

5. 龙龟 Lóng Guī Dragon Tortoise — 5 Water

Lóng Guī brings good *feng shui*; promotes safety and longevity; its head bestows blessings and wealth; its tail wards off evil.

The 8 Immortals 八仙

Stories of humans who had become immortals date back to the Han Dynasty and before. However, members of the group known as the Eight Immortals 八仙 Bā Xiān were mostly born in the Tang or Song Dynasty and are thought to live on five islands in the Bóhǎi Sea 渤海. They are often depicted in art crossing the sea from the city of Pénglái 蓬莱 on Bóhǎi's southern shore. Each of them has a power tool 法器 Fǎ Qì and these tools or ritual implements are collectively known as the Covert Eight Immortals (暗八仙 Àn Bā Xiān). Besides their implements there are other things which identify some of them. Zhāng Guǒ Laǒ 张果老 rides a donkey and carries a peach of immortality; Lǚ Dòng Bīn 吕洞宾 carries a fly whisk; Tiě Guǎi Lǐ 铁拐李 holds an iron crutch; Hé Xiān Gū 何仙姑 is a woman; she holds a fly whisk and is accompanied by a phoenix; and the sex of Lán Cǎi Hé 蓝采和 is uncertain. He wears one shoe, carries a pick or hoe and is accompanied by a Crane.

Tarot Mahjong 塔罗麻将牌 — The 8 Immortals 八仙过海

25 ◇ **The 8 Immortals 八仙过海**

3. 曹国舅 Cáo Guó Jiù — **3** Earth 土

过 cross

富 Wealth
(Royal Uncle Cao)

艮 Gèn

---

25 ♥ **The 8 Immortals 八仙过海**

4. 韩湘子 Hán Xiāng Zǐ — **5** Water 水

海 the sea.

少 Youth

坎 Kǎn

---

35 🏯 Great Wall Guard Tower 长城隘口碉楼 **The 8 Immortals 八仙过海**

5. 铁拐李 Tiě Guǎi Lǐ — **4** Metal 金

各 Each

贱 Humility
(Iron Crutch Li)

兑 Duì

---

25 ♣ **The 8 Immortals 八仙过海**

6. 汉钟离 Hàn Zhōng Lí — **2** Fire 火

显 shows

贵 Nobility
(Zhōng Lí Quán 钟离权)

离 Lí

The 8 Ritual Implements of the Immortals 八仙法器

Each of the Eight Immortals has a power tool (法器 Fǎ Qì) and these tools or ritual implements are collectively known as the Covert Eight Immortals (暗八仙 Àn Bā Xiān) or the Eight Ritual Implements of the Immortals (八仙法器 Bā Xiān Fǎ Qì). The ritual implements are a fish drum, a sword, a pair of jade tablets, a flute, a gourd, a fan, a lotus flower and a flower basket. The drawings are based on wood carvings of these implements in Yán Qìng Guān 延庆观 Daoist Temple in Kaifeng City, Henan Province.

Tarot Mahjong 塔罗麻将牌
Eight Ritual Implements of the Immortals 八仙法器

26 ♠

1. 鱼鼓 Yú Gǔ (Fish Drum) — **1** 木 Wood

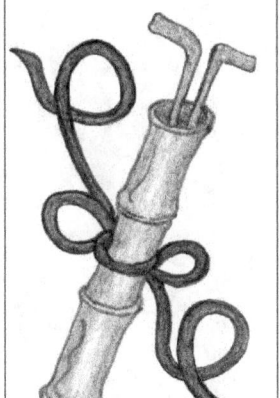

Fish Drum carried by Zhāng Guǒ Lǎo 张果老

震 Zhèn

---

34 🏯 Great Wall Guard Tower 长城岗楼

2. 宝剑 Bǎo Jiàn (Sword) — **4** 金 Metal

Sword carried by Lǚ Dòng Bīn 吕洞宾

乾 Qián

---

26 ♦

3. 玉板 Yù Bǎn (Jade Tablets) — **3** 土 Earth

Jade Tablets carried by Cáo Guó Jiù 曹国舅

艮 Gèn

---

26 ♥

4. 笛子 Dí Zǐ (Flute) — **5** 水 Water

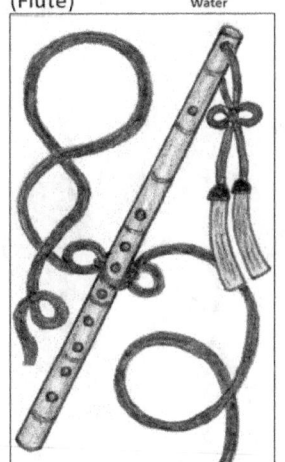

Flute carried by Hán Xiāng Zǐ 韩湘子

坎 Kǎn

Tarot Mahjong 塔罗麻将牌
Eight Ritual Implements of the Immortals 八仙法器

36 Great Wall Guard Tower

5. 葫芦 Hú Lu (Gourd) — 4 金 Metal

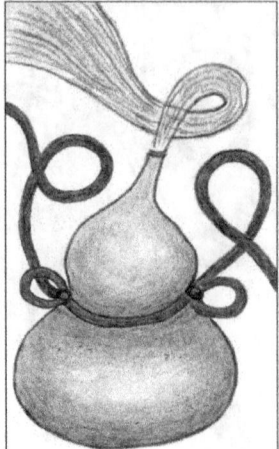

Gourd carried by Tiě Guǎi Lǐ
铁拐李

兑 Duì

Tarot Mahjong 塔罗麻将牌
Eight Ritual Implements of the Immortals 八仙法器

26 ♣

6. 扇子 Shàn Zǐ (Fan) — 2 火 Fire

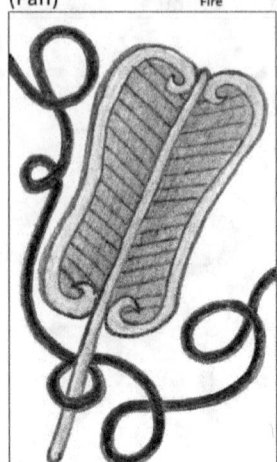

Fan carried by Hàn Zhōng Lí
汉钟离

离 Lí

Tarot Mahjong 塔罗麻将牌
Eight Ritual Implements of the Immortals 八仙法器

28 ♦

7. 荷花 Hé Huā (Lotus) — 3 土 Earth

Lotus carried by Hé Xiān Gū
何仙姑

坤 Kūn

Tarot Mahjong 塔罗麻将牌
Eight Ritual Implements of the Immortals 八仙法器

28 ♠

8. 花篮 Huā Lán (Flower Basket) — 1 木 Wood

Flower Basket carried by Lán Cǎi Hé
蓝采和

巽 Xùn

Chinese Chess 中国象棋

The earliest extant Chinese chess pieces date from the Song Dynasty (960-1279) while the earliest written reference to the game is by the Tang Dynasty minister Niú Sēng Rú 牛僧孺 (779-847). Xiàngqí is derived from the Indian game of Chaturanga which developed in the 6th century. Chaturanga means "having four parts" or "four arms" which are the elephants, chariots, cavalry and infantry. Elephants appear both in the Indian game and in the Chinese game which is called "Chinese Elephant Chess". Elephants were rarely employed in Chinese armies although two elephants did fight in an army of the Western Wei Dynasty 西魏 in 554 AD. A chess piece called Prime Minister, which has the same pronunciation as elephant, is used opposite the elephants in the Chinese game.

Qín Dynasty General; Builder of the Great Wall
秦朝将军，修建长城

General who led rebel forces against the Qín Dynasty
领导叛军反抗秦朝

②   Chinese Chess 中国象棋

侍卫 Shì Wèi **1** 木 Wood

Shì

Tarot
♠11
♣11
♢11
♡11

Bodyguard of a General or King
帝王或将军左右卫护的武官

②   Chinese Chess 中国象棋

侍卫 Shì Wèi **1** 木 Wood

Shì

Tarot
♠11
♣11
♢11
♡11

Bodyguard of a General or King
帝王或将军左右卫护的武官

③ Chinese Chess 中国象棋

西魏 Western Wèi
535-557 AD **3** 土 Earth

Xiàng

Tarot
🂠15

Western Wèi Dynasty used elephants in army in 554 AD
公元554年，北魏战争中使用大象

③  Chinese Chess 中国象棋

李斯 Lǐ Sī
(ca. 280 BC-208 BC)

Xiàng

Prime Minister of Qín
秦朝丞相

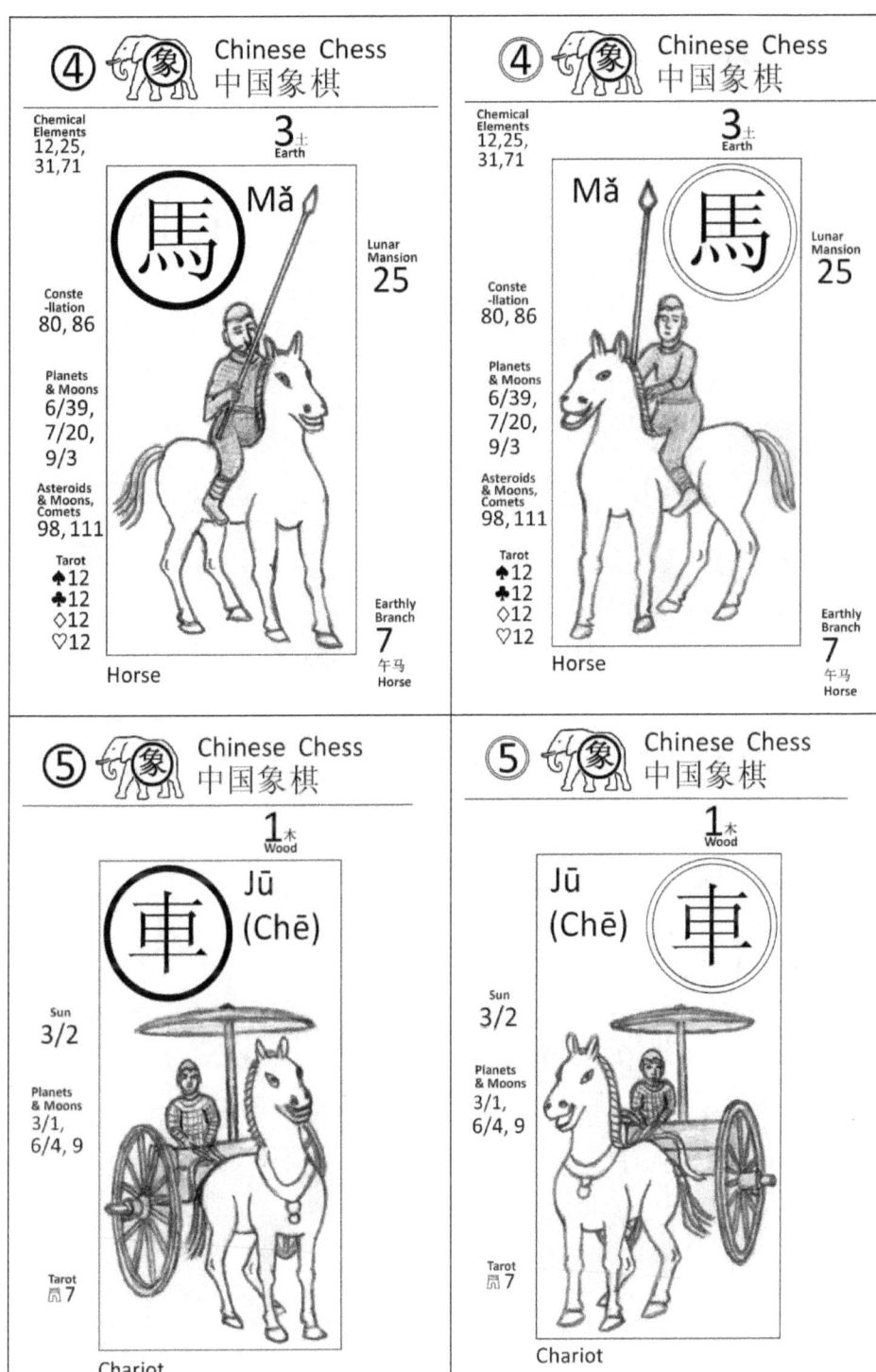

⑥  Chinese Chess 中国象棋	⑥  Chinese Chess 中国象棋
4 金 Metal	4 金 Metal
Pào  Heavenly Stem 7 庚 Gēng	Pào  Heavenly Stem 7 庚 Gēng
Cannon	Cannon
⑦   Chinese Chess 中国象棋	⑦  Chinese Chess 中国象棋
4 金 Metal	4 金 Metal
Zú 	Bīng 
Soldier	Soldier

Index

Abbreviations
(60Y) 60 Year Cycle; (E) Eris;
(H) Haumea; (J) Jupiter; (M) Mars;
(N) Neptune; (O) Orcus; (P) Pluto;
(Q) Quaoar; (S) Saturn; (SL) Salacia;
(U) Uranus; (V) Varda.

	Page
Achelous, River God	182
Achelous, River God	182
Achilles (Asteroid) Greek Warrior	257
Achilles and King Priam	256
Achilles murders Troilus	259
Actaea (SL), Nymph of sea shore	221
Actinium (Element)	135
Actis, Son of Helios	135
Adonis	181
Adonis (Asteroid) God of Crops	228
Adrastea (J)	171, 176
Aeda, Irish Giant	196
Aegaeon (S) Giant	190
Aegina (Asteroid) River Nymph	236
Aegir (S) Norse Sea God	201
Aegis worn by Athena	241
Aeneas (Asteroid) and Tiber River	256
Aeneas carries his father Anchises	259
Aether (Ether) (Five Elements)	4
Agamemnon (Asteroid)	257
Aglauros and Herse	179
Air (Five Element)	3
Aitne (J) Mt Etna	179
Ajax (Asteroid) son of Telamon	259
Akhenaten, Egyptian Pharaoh	224
Alauda (Asteroid) Latin for Lark	252
Albertus Magnus	121
Albiorix (S) (Teutates)	196
Alchemist (Magician) in Tarot	275
Alexhelios (Alexander Helios)	244
Alioth (Northern Dipper)	104
Alkaid (Northern Dipper)	105
Allied Intervention Siberia (60Y)	74
Aluminium (Element)	116
Amalthea (J)	171
American Platinum Eagle Coin	132
Americium (Element)	136
Amerigo Vespucci	136
Americas (North and South)	135
Amor (Asteroid) God of Love	229
Amphitrite (Asteroid)	235
Amphitrite (Sea Goddess)	190
Ananke (Anance) (J)	179
Anchises (Asteroid) with Aeneas	259
Andromeda (Constellation)	142
Animals of the 5 Directions	11
Anser, the Goose	160
Anteros statue, London	116
Anthe (S)	191
Antilochus (Asteroid)	255
Antimony (Element)	125
Antlia (Constellation)	151
Aoede (J) (Muse)	176, 178, 181
Aphrodite (Venus)	168
Apollo	127
Apollo (Asteroid)	227
Apus (Constellation)	157
Aquarius (Constellation)	162
Aquarius (Zodiac)	54
Aquila (Constellation)	160
Ara (Constellation)	157
Arche (J) (Muse)	175, 185
Ares, Greek War God	169
Argo (Ship)	147, 149
Argon (Element)	117
Argonauts	257
Argos guards the cow Io	235
Ariel (U) Sylph of Air	209
Aries (Constellation)	143
Aries (Zodiac)	52
Army Defeating Star (N Dipper)	106
Arsenic (Element)	121
Artemis	127

Ascella (Southern Dipper)	107	Běihǎi Park, White Pagoda	272
Asclepius	158	Běijīng, Great Ancient Capital	288
Aspasia (Asteroid) Wife of Pericles	236	Běijīng-Hankou Railway (60Y)	72
Astatine (Element)	134	Běijīng-Paris Car Race (60Y)	72
Asteroids and Moons, Comets	225	Belinda (U) lost her lock of hair	208
Astraea (Asteroid) Star Maiden	235	Bell Dragon, Pú Láo	111
Aten (Asteroid) Egyptian Sun God	225	Bergelmir (S) hides from flood	199
Athena	190	Berkelium (Element)	137
Athena and Owl coin	124	Beryl (Gemstone)	113
Atlas (S)	189	Beryllium (Element)	113
Atlas (Titan)	118	Bestla (S) Norse Frost Giantess	202
Atlas' Daughters	179	Bettina (Asteroid) von Rothschild	250
Auriga (Constellation)	147	Bì Àn Dragon Dog (Lunar M)	30
Ausonia, Campania (Asteroid)	232	Bì Àn Dragon Dog (9 Dragon Sons)	110
Autonoe (J)	187	Bì Àn Tarot, Symbol of Justice	278
Autumn (Seasons)	6	Bì Xi (Bà Xià) (9 Dragon Sons)	111
Autumn Begins (Solar Term)	36	Bì Xì (Fù Xī) (9 Dragon Sons)	111
Autumn Equinox (Solar Term)	36	Biàn Shì, Queen of Wèi	266
Autumn Tarot-Mahjong	290, 291	Bianca (U) Cassio's Mistress	206
Auxo (Horae)	174	Bianca (U) Daughter of Baptista	205
Azure Dragon of the East	11	Birdman, Easter Island (Rapa Nui)	221
Azure Emperor	7	Bismuth (Element)	133
Bā Guà (8 Trigrams)	79-86	Black Emperor (Mercury)	8
Bā Xià (Gōng Fù) (9 Dragon Sons)	112	Black Star (Northern Dipper)	104
Bà Xià (Bì Xì) (9 Dragon Sons)	111	Black Tiger Spring, Jínán,	282
Bac Le (Běi Lí) Ambush (60Y)	67	Black Warrior of the North	12
Bacchus (Asteroid) Wine God	225	Bodyguard of General or King	309
Badger (Lunar Mansion)	25	Bodyguard of King	264, 268, 272, 283
Bái Huā, White Flower	272	Bohr, Niels (Danish Physicist)	139
Balance (Libra)	53	Bohrium (Element)	139
Balance, Tarot Symbol of Justice	277	Book of Changes (Yì Jīng)	86-102
Balder's Funeral Ship	197	Boötes (Constellation)	155
Bamberga (Asteroid) Bamberg	238	Boron (Element)	114
Bamboo	291	Boxer Protocol (60Y)	71
Bank of China	277	Boxer Uprising (60Y)	70
Bāo Zhěng judges in the afterlife	280	Bridge (Flower Cards)	43
Barium (Element)	126	British invade Tibet 1903 (60Y)	71
Bat (Lunar Mansion)	27	Bromine (Element)	121
Bat represents Fú Xīng	295, 297	Bronze Coins (Heavenly Stems)	57
Battle of Mukden (Shěnyáng)	72, 278	Bull (Taurus)	52
Bear (Martial Arts Bā Guà)	87	Bush Clover (Flower Cards)	45
Bebhionn (S) Irish Giantess	196	Butterfly (Flower Cards)	44

Cadmium (Element)	124	Cerberus (Asteroid)	226
Cadmus	124	Cerberus kills Pirithous	184
Caelum (Constellation)	145	Ceres (Minor Planet)	169
Caesium (Element)	126	Ceres (Roman Grain Goddess)	127
Cái Shén, God of Wealth	296	Cerium (Element)	127
Calcium (Element)	117	Ceto(Asteroid) Sea Goddess	262
Cale (Kale) Goddess of Beauty	179	Cetus (Constellation)	142
Caliban (U) Savage Slave	210	Chaldene (J) Mother of Solymos	180
California (Asteroid)	229	Chamaeleon (Constellation)	152
California Republic	136	Cháng'ān, Great Ancient Capital	288
California, University of	136	Cháng É (Moon Goddess)	21
Californium (Element)	137	Cháng É, Tarot Moon	280
Calligraphy, Mahjong 4 Pastimes	292	Cháng Gēng (Venus)	8
Callirrhoe (J) Achelous' Daughter	182	Change of Seasons	6
Callisto (J)	172	Cháo Fēng (Roof Corner Dragon)	111
Calypso (S)	192	Chāo Yǒng cruiser launched (60Y)	66
Camelo Pardalis (Constellation)	150	Charcoal	114
Camilla (Asteroid)	253	Chariklo (Asteroid) wife of Chiron	260
Campania, Italy (Ausonia)	231	Chariot	277, 310
Cancer (Constellation)	149	Charites, Graces	176, 179, 183,185,248
Cancer (Zodiac)	52	Charon (P) Boatman of the Dead	219
Canes Venatici (Constellation)	154	Chemical Elements	112-140
Cāng Dì (Jupiter)	7	Chén Xīng (Mercury)	8, 22
Canis Major (Constellation)	147	Chéng Hǎi Lóu Temple burnt (60Y)	71
Canis Minor (Constellation)	149	Cherry Blossoms (Flower Cds)	41
Cannon (Chinese Chess)	311	Chess, Mahjong Four Pastimes	292
Cáo Cāo, King of Wèi	266	Chī Wěn (Rooftop Dragon)	111
Cáo Guó Jiù (Wealth)	304	Chicken (Later Heaven)	84
Cape Town, South Africa	146	China Merchant Steamship (60Y)	64
Capricorn (Zodiac)	54	Chinese Chess	308
Capricornus (Constellation)	161	Chinese Earth God Tǔ Dì Shén	168
Carbon (Element)	114	Chinese Eastern Railway (60Y)	70
Carina (Constellation)	150	Chiron (Asteroid and Comet)	260
Carme (J) Cretan Harvest Nymph	182	Chiron (Centaur)	154
Carpo (J) (Horae)	174	Chlorine (Element)	117
Cassandra clings to the Palladium	124	Chloris (Asteroid) Flower Nymph	239
Cassia Tree on the Moon	21	Chloris, Flower Nymph (Elements)	117
Cassiopeia (Constellation)	142	Chóng Zhēn, Ming Emperor	278
Cat, Four Mahjong Animals	294	Chromium (Element)	118
Centaurus (Constellation)	154	Chrysanthemum (Flower Cards)	47
Centipede, Four Mahjong Animals	294	Chrysanthemum, Mahjong	290, 291
Cepheus (Constellation)	143	Chryseis, Daughter of Chryses	256

Chryses, Apollo's Priest	256	Curium (Element)	136
Chǔ Zhuāng Wáng, King of Chǔ	274	Cybele (Asteroid)	253
Cí Xǐ returns from Xī'ān (60Y)	71	Cygnus (Constellation)	160
Circe (Asteroid)	238	Cyllene (J) Nymph	184
Circinus (Constellation)	154	Dactyl (Rhea's fingers)	245
Civil Star of Mystery (N Dipper)	105	Daedalus (Asteroid)	227
Classic of Changes (Yì Jīng)	87-103	Dali 3 Pagodas, Yunnan	78
Clear Bright (Solar Term)	34	Dāndōng, Tiger Mt Great Wall	262
Cleoselene (Cleopatra Selene)	244	Daoist Mountains	9-10
Cobalt (Element)	119	Daphne (Asteroid) Laurel Tree	242
Cockerel, Four Mahjong Animals	294	Daphnis (S) Sicilian Musician	189
Cold Dew (Solar Term)	37	Darmstadt, Germany	140
Columba (Constellation)	147	Darmstadtium (Element)	140
Coma Berenices (Constellation)	153	Davida (Asteroid) David Peck Todd	251
Communist Party Congress (60Y)	76	Death (Tarot, Battle of Mukden)	278
Concordia (Asteroid)	238	Deer (Flower Cards)	48
Confucius (Kǒng Zǐ)	268	Deer (Lunar Mansion)	31
		Deer represents Lù Xīng	295,297
Constellations	141-163	Deimos (M)	169
Copenhagen, Denmark	130	Delphi, Oracular Shrine	194
Copernicium (Element)	140	Delphinus (Constellation)	161
Copernicus, Nicolaus	140	Dembowska (Asteroid)	246
Copper (Element)	120	Dembowski, Baron Ercole	246
Cordelia (U) Daughter of King Lear	205	Demeter	169
Corona Austrina (Constellation)	158	Desdemona (U) Wife of Othello	206
Corona Borealis (Constellation)	156	Despina (N) Daughter of Poseidon	214
Corvus (Constellation)	153	Devil, Tarot, Mó Zhàng (Mara)	279
Cow (Later Heaven)	83	Dia (J) Wife of Ixion	174
Crab (Cancer)	52	Diamonds	114
Crane (Flower Cards)	39	Dìng Yuǎn Cruiser scuttled (60Y)	69
Crane representa Shòu Xīng	296,298	Diomedes (Asteroid)	256
Crater (Constellation)	152	Dione (Asteroid) Dodona Oracle	251
Cressida (U) Lady loved by Troilus	206	Dione (S) Goddess at Dodona	193
Crommelin's Comet	260	Dionysus	178, 182, 187
Cronos (Saturn) and Rhea	188,193	Diotima (Asteroid) and Socrates	247
Crotus (Centaur)	159	Directions depicted as Animals	11-12
Crow (Lunar Mansion)	29	Dodona Oracular Shrine	193, 251
Crux (Constellation)	153	Dog (Earthly Branch)	61
Cuckoo (Flower Cards)	42	Dog (Later Heaven)	84
Cupid (U) Boy at Timon's Banquet	208	Dog (Lunar Mansion)	29
Cupid marries Psyche	246	Dorado (Constellation)	146
Curie, Marie and Pierre	136	Dǒu Mǔ (North Star Goddess)	104,279

Doyōbi (Saturday)	19	Emma (Asteroid)	246
Draco (Constellation)	155	Emperor Táng Gāo Zōng (Tarot)	276
Dragon and Phoenix coinage	271	Emperors (Five)	7-8
Dragon (Earthly Branch)	59	Empress Wǔ Zé Tiān (Tarot)	276
Dragon (Later Heaven)	84	Enceladus (S) Giant	179, 192
Dragon (Lunar Mansion)	25	Encke's Comet	230
Dragon (Martial Arts)	86	Encke, Johann Franz (Astronomer)	230
Dragon Dog, Bì Àn (Lunar M)	30	Eosphoros, God of Dawn	116
Dragon Dog, Bì Àn (9 Dragon Sons)	110	Epimetheus (S) and Pandora	190
Dragon Door Guard, Jiāo Tú	110	Equuleus (Constellation)	161
Dragon Lion, Suān Ní	110	Erbium (Element)	129
Dragon, Old Dragon Head 老龙头	280	Eridanus (Constellation)	144
Dragon Tortoise Lóng Guī	302	Erinome (J)	181
Dubhe the Bear (N Dipper)	103	Eris (Dwarf Planet)	224
Dubnium (Element)	139	Eros (Asteroid) God of Love	227
Dysnomia (E) Daughter of Eris	224	Erriapus (S) God of Green plants	196
Dysprosium (Element)	129	Esus prunes a tree	198
Earth (Earlier Heaven)	81	Etna, Mount, Italy	178
Earth (Five Elements)	2, 4	Euanthe (J) Mother of Charites	176
Earth (Gaea) (Planet)	167	Eugenia (Asteroid) Empress	239
Earth (Later Heaven)	83	Eukelade (J) (Muse)	184
Earlier Heaven Trigrams	80-82	Eunomia (Asteroid)	237
Earth, Suit in Tarot	271-274	Euphrosyne (Asteroid)	250
Earth God Tǔ Dì Shén	286	Euporie (J) (Horae)	175, 177
Earthly Branches (Twelve)	58-61	Europa (Asteroid) with Bull	248
Easter Island, Makemake	221	Europa (J)	172
Echidna (Asteroid) Typhon's wife	262	Europa and the Bull	128
Egeria (Asteroid) Water Nymph	235	Europium (Element)	128
Egyptian black eye paint	124	Eurydome (J) Mother of Charites	183
Eight Immortals (Bā Xiān)	303-305	Euterpe (Asteroid) (Muse)	231
Eight Ritual Implements	305	Everest, Mt (Zhū Mù Lǎng Mǎ)	114
Eight Trigrams (Bā Guà)	79-87	Fā Cái, Making a Fortune	268
Einstein, Albert	137	Fá Chāng Machine Factory (60Y)	63
Einsteinium (Element)	137	Fǎ Xiǎn walks to India	282
Ejnar Hertzsprung (Astronomer)	226	Fan carried by Hàn Zhōng Lí	307
Elephant, Western Wèi Dynasty	308	Fán Jī, Queen of Chǔ	274
English Days of the Week	13-15	Fāng Jǔ-zàn (60Y)	63
Elara (J)	174	Farbauti (S) Storm Giant	201
Elder Zhāng Guǒ (Zhāng Guǒ Lǎo)	302	Farmer, 4 Mahjong Professions	293
Elektra (Asteroid)	248	Fèngtiān: see Shěnyáng	68
Elements (Five)	1-4	Fèngtiān Medical College (60Y)	69
Elephant (Chinese Chess)	309	Fenrir (S) Monstrous Wolf	202

Ferdinand (U) Son of King Alonso	212	Four Phenomena	76-79
Fermi, Enrico	137	Four Professions	293
Fermium (Element)	137	Four Seasons	290
Fire (Earlier Heaven)	82	Fox (Lunar Mansion)	26
Fire (Five Elements)	1, 3	France (Coin)	134
Fire (Later Heaven)	83	Francisco (U) Attendant to Alonso	210
Fire Planet (Mars)	21	Francium (Element)	134
Fire, Tarot Suit	267-270	Freya (Vanadis), Norse Goddess	118
First Communist Congress (60Y)	76	Friday (Frigg)	14
First Zhílì-Fèngtiān War (60Y)	76	Frigg (Goddess)	14
Fish and Yāo Jī Bird	266	Frog (Flower Cards)	49
Fish Drum held by Zhāng Guǒ Lǎo	306	Frost Falls (Solar Term)	37
Fish represents Cái Shén	296, 298	Fù Hǎo, Female General	265
Fisherman, 4 Mahjong Professions	293	Fǔ Jìn Mén Gate Shěnyáng	272
Five Animals of Fortune Stars	297	Fú Sāng Tree in East (10 Suns)	164
Five Auspicious Animals	301-302	Fù Xī (Bì Xì) (9 Dragon Sons)	112
Five Directions (Animals)	11-12	Fú Xī (Emperor, Sun God)	7, 23
Five Elements	1-4	Fú Xīng, Star of Good Fortune	295
Five Emperors	7-8	Gadolin, Johan (Finnish Chemist)	128
Five Great Ancient Capitals	287-289	Gadolinium (Element)	128
Five Stars of Good Fortune	295	Gaea, Earth (Planet)	168
Five Symbols of Prosperity	299	Galatea (N) Lady of White Foam	214
Flame and Sun (Heavenly Stem)	55	Gallia (Asteroid) Roman France	243
Flammario (Asteroid) Flammarion	240	Gallium (Element)	120
Flood Dragon (Lunar Mansion)	25	Gān Fū Ren, Queen of Shǔ	270
Flora (Asteroid) Flower Goddess	229	Gān Yuān abyss (10 Suns)	165,166
Flower Basket held by Lán Cǎi Hé	307	Ganymede (J)	172
Flower Cards (Hanafuda)	39-51	Ganymede (Rosalind)	207
Flower Viewing Curtain	41	Gāo Lóu Zhài, Battle of (60Y)	61
Fluoride	114	Geiger Counter	133
Fluorine (Element)	115	Gemini (Constellation)	148
Flute carried by Hán Xiāng Zǐ	306	Gemini (Zodiac)	52
Fool, Major Arcana	275	Geographos (Asteroid) Strabo	226
Fornax (Constellation)	144	German Coat of Arms on coin	120
Fornjot (S)	204	Germania (Asteroid)	247
Fortuna (Asteroid)	234	Germanium (Element)	120
Fortune Cat Zhāo Cái Māo	302	Germans land at Qīng Dǎo (60Y)	70
Four Flowers (Mahjong)	291	Getsuyōbi (Monday)	18
Four Greek Elements	3-4	Gibbon (Lunar Mansion)	30
Four Great Heavenly Kings	285	Giraffe sent to China from Africa	284
Four Mahjong Animals	294	Gjalp	197
Four Pastimes	292	Gnome of the Earth	4

Goat (Earthly Branch)	60	Hán Fēi Zǐ Legalist Philosopher	269
Goat (Later Heaven)	83	Hán Xiāng Zǐ (Youth)	304
Goat (Lunar Mansion)	30	Hàn Yáng Iron & Steel Wks (60Y)	68
Goddess of North Star Dǒu Mǔ	103	Hàn Zhōng Lí (Nobility)	304
Goddess of Planet Venus	22	Hanafuda (Flower Cards)	38-50
Gold (Element)	132	Hanged Man (Tarot, Chóng Zhēn)	278
Golden Money Toad, Jīn Chán	301	Harbin Songhua River	79
Golden Mother of Lake Yáo	276	Harmonia (Asteroid)	230
Golden Treasure Ingot	295, 300	Harpalyke (J)	178
Gōng Fù (Bā Xià) Water Dragon	112	Hassium (Element)	139
Gorgoneion (Gorgon Medusa)	241	Hati (S) Norse Wolf chases Moon	200
Gǒu Chén, (Northern Dipper)	106	Haumea (Dwarf Planet)	221
Gourd carried by Tiě Guǎi Lǐ	307	Hé Xiān Gū (Femininity)	305
Gourmet Dragon, Tāo Tiè	110	Heaven (Earlier Heaven)	80
Graces, Charites	176, 179,183,185,248	Heaven (Later Heaven)	83
Grain in Ear (Solar Term)	35	Heavenly Kings	284-286
Grain Rain (Solar Term)	34	Heavenly Stems	55-57
Grand Canal	272	Hebe (Asteroid) Goddess of Youth	233
Graphite (Carbon)	114	Hecatebolus (Southern Dipper)	108
Great Cold (Solar Term)	38	Hector (Asteroid)	257
Great Gate Star (Northern Dipper)	104	Hegemone (J) Leader of Charites	185
Great Heat (Solar Term)	35	Hèlán Shān Sun God	24
Great Heavenly Emperor (N Dipp)	106	Helene (S) Wife of Menelaus	193
Great Primal Beginning (Tài Jí)	78	Helen's Husband Menelaus	257
Great Purple Emperor (N Dipp)	106	Helike (Helice) (J)	171, 176
Great Tea Race (60Y)	62	Helio (Asteroid) Helios, Sun God	252
Great Wall at Jiā Yù Guān	272	Heliopolis, Egypt	135
Great Wall at Tiger Mountain	262	Helios (Greek Sun God)	113, 167, 187
Great Wall meets the Sea	280	Helium (Element)	113
Greedy Wolf Star (N Dipper)	104	Héng Shān, Húnán (Mt Héng)	9
Green Star (Northern Dipper)	105	Héng Shān, Shānxī (Mt Héng)	10
Greip (S)	198	Hera (Asteroid) Wife of Zeus	238
Grigg-Mellish (Comet)	261	Hercules (Constellation)	158
Grus (Constellation)	162	Hercules helped by Telamon	255
Guì Shù (Cassia Tree on Moon)	21	Herculina (Asteroid) Hercules	243
Gunpowder (Mongol Soldier)	116	Hermes (Mercury)	167
Hǎi Qí, Hǎi Róng naval ships (60Y)	73	Hermione (Asteroid)	253
Hafnia (Copenhagen, Denmark)	130	Hermippe (J)	178
Hafnium (Element)	130	Hermit, Poet Lín Bū	277
Halimede (N) Lady of the Brine	216	Herse (J) and Aglauros	179
Halley's Comet	261	Hertzsprung, Ejnar and Ivar	228
Hàn Dynasty War Chariot	276	Hesse (German State)	139

Hestia (Asteroid)	234	Immortal Woman Hé (Hé Xiān Gū)	304
Hexagrams, Sixty four	86-102	Insects Awakening (Solar Term)	33
Hibiscus and Sun Bird (10 Suns)	164	Indium (Element)	125
Hidalgo (Asteroid)	260	Indus (Constellation)	162
High Priest Zhāng Dào Líng (Tarot)	276	Interamnia (Asteroid) Italian City	248
High Priestess Xī Wáng Mǔ	276	Io (J) Lover of Zeus	171
Hi'iaka (H) Hula Dancing Goddess	222	Io (Asteroid)	237
Hilda (Asteroid) Battle Maiden	255	Iocaste (Jocasta) (J)	177
Himalia (J)	173	Iodine (Element)	126
Hindenburg Airship	113	Irene (Eirene) (Asteroid)	236
Hispania (Asteroid) Spain	244	Iridium (Element)	132
Holmium (Element)	129	Iris (Flower Cards)	43
Hong Kong College Medicine (60Y)	67	Iris, Rainbow Goddess	132
Hong Kong Golf Club (60Y)	68	Iris (Asteroid) Rainbow Goddess	232
Horae (Hours, Seasons)	174-177	Iron (Element)	119
Horologium (Constellation)	144	Iron Crutch Lǐ (Tiě Guǎi Lǐ)	303
Horse	265, 269, 273, 283, 310	Isis (Asteroid) Egyptian Goddess	234
Horse (Earthly Branch)	60	Isonoe (J)	185
Horse (Later Heaven)	83	Ivar (Asteroid) Brother of Ejnar	228
Horse (Lunar Mansion)	31	Ixion (Dwarf Planet)	220
Hòu Yì shoots down 9 suns	280	Jade Star (Northern Dipper)	105
Hù Sān Niáng Swordswoman	263	Jade Buddha Temple, (60Y)	66
Huá Shān, Shaǎnxi (Mt Huá)	10	Jade Rabbit on Moon	21
Huáng Dì, Yellow Emperor	8	Jade Tablets held by Cáo Guó Jiù	306
Huì Shēn returns from sea voyage	282	Janus (S) God of Doors and Gates	190
Hungaria (Asteroid) Hungary	229	Japan captures Port Arthur (60Y)	71
Huǒ Xīng (Mars)	7, 21	Jarnsaxa (Giantess)	198
Huya (Juyá) Rain God of Wayúu	218	Jī lóng to Táiběi Railway (60Y)	69
Hydra (Constellation)	152	Jiā Yù Guān Great Wall	272
Hydra (P) Nine-headed Serpent	220	Jiàn Zhēn sails to Japan	283
Hydrogen (Element)	113	Jiāo Lóng, Flood Dragon	24
Hydrus (Constellation)	143	Jiāo Tú at door (9 Dragon Sons)	110
Hygiea (Asteroid)	250	Jiāo Zhōu Bay, Qīngdǎo (60Y)	70
Hyperion (S) Father of Sun, Moon	194	Jīn Chán Golden Money Toad	301
Hypnos (Sleep) and Pasithee	183	Jínán City, Black Tiger Spring	282
Hyrrokkin (S) Norse Giantess	198	Jìng Yuǎn Cruiser launched (60Y)	67
Iapetus (S) Father of Atlas	194	Jocasta (Iocaste)	176
Icarus (Asteroid) falls into the sea	226	John III Sobieski of Poland	158
Ida (Asteroid) Nurse of baby Zeus	244	Jù Mén (Northern Dipper)	104
Ijiraq (S) Inuit Spirit	195	Judgement (Tarot) Bāo Zhěng	280
Ili (Yī Lí) XīnJiāng, Russians (60Y)	63	Julia, Saint of Corsica (Asteroid)	235
Ilmarë, (V) Varda's handmaiden	222	Juliet (U) Lover of Claudio	207

Juliet (U) Lover of Romeo	206	Lecoq de Boisbaudran, Paul-Emile	128
Juno (Hera) (Asteroid)	237	Léi Fēng Pagoda, Hángzhōu	279
Jupiter (Jovis, Thursday)	16	Leda (J) and the Swan	173
Jupiter (Mù Xīng, Suì Xīng)	7, 22	Leda (Asteroid) with Swan	240
Jupiter (Zeus) (Planet)	170	Leo (Constellation)	151
Juyá (Huya) Rain God of Wayúu	217	Leo (Zodiac)	53
Kāifēng, Great Ancient Capital	289	Leo Minor (Constellation)	151
Kale (Cale) (J) Goddess of Beauty	179	Leopard (Lunar Mansion)	26
Kallichore (J)	182	Lepus (Constellation)	146
Kalliope (Asteroid) Muse	245	Lǐ Sī Prime Minister of Qín	309
Kalyke (J)	182	Lián Zhēn, Star (Northern Dipper)	105
Kalypso (Calypso) (Asteroid)	236	Libra (Constellation)	155
Kari (S) Ruler of the North Wind	203	Libra (Zodiac)	53
Kaus Borealis (Southern Dipper)	107	Light Snow (Solar Term)	37
Kayōbi (Tuesday)	18	Lightning (Earlier Heaven)	80
Kerberos (Cerberus) (P)	220	Lightning (Flower Cards)	50
King, Tarot	265, 269, 273, 284	Lightning (Later Heaven)	83
Kinyōbi (Friday)	19	Limit of Heat (Solar Term)	36
Kiviuq (S) Inuit Hero	194	Lín Bū, Hermit, Poet	277
Kleopatra (Cleopatra) (Asteroid)	243	Lín Chōng	265
Knight	265, 269, 273, 283	Linus Son of Calliope	245
Knight in Chinese Chess	309	Lion (Leo)	53
Kobold (German Goblin)	119	Lion (Martial Arts Bā Guà)	85
Kǒng Zǐ, Confucius	268	Lion, Strength in Tarot	276
Kore (J)	184	Literature Dragon, Fù Xī (Bì Xì)	112
Krypton (Element)	121	Lithium (Element)	113
Kūn Lún Mountains	276	Little Cold (Solar Term)	38
Lacerta (Constellation)	163	Little Milk Dipper	106-107
Laetitia (Asteroid) Goddess of Joy	242	Liú Bèi, King of Shǔ	270
Lamp Flame (Heavenly Stem)	56	Liú Táng	265
Lán Cǎi Hé (Poverty)	305	Loge (Logi) (S) Fire Giant	203
Lanthanum (Element)	127	Lóng (Dragon) (Lunar Mansion)	24
Lǎo Lóng Tóu (Old Dragon Head)	281	Lóng Guī, Dragon Tortoise	302
Lǎo Zǐ, Daoist Philosopher	267	Loreley (Asteroid)	249
Laomedeia (N) Lady who leads	216	Lotus Blossom, Mahjong Seasons	290
Larissa (N) Thessalian Nymph	214	Lotus carried by Hé Xiān Gū	307
Lascaux Cave Paintings	119	Lovers Zhī Nǚ, & Niú Láng (Tarot)	277
Lasting Prosperity Star (N Dipper)	105	Lù Cún, Star (Northern Dipper)	105
Later Heaven Bā Guà	82-84	Lǚ Dòng Bīn (Masculinity)	303
Lawrence, Ernest Orlando	138	Lù Xīng, Star of Emolument	295
Lawrencium (Element)	138	Luminaries, Chinese	20-24
Lead (Element)	133	Luna (Monday)	16

Luna (Selene) Moon	168	Menelaus (Asteroid)	258
Lunar Mansions, China	24-31	Méng Tián, Qín Dynasty General	308
Luòyáng, Great Ancient Capital	288	Mèng Zǐ, Mencius, Confucian Phil.	268
Lupus (Constellation)	156	Menoetius, Father of Patroclus	258
Lutetia (Asteroid) Paris	233	Mensa (Constellation)	146
Lutetia (Paris)	130	Merak (Northern Dipper)	103
Lutetium (Element)	130	Mercury (Element)	132
Lycaon (J) Wolf, Son of Cyllene	184	Mercury (Wednesday)	16
Lynx (Constellation)	149	Mercury (Hermes) (Planet)	167
Lyra (Constellation)	159	Mercury (Shuǐ Xīng, Chén Xīng)	8, 22
Lysithea (J) (Semele, Thyone)	173, 178	Metal (Five Elements)	1-4
Mab (U) Fairies' Midwife	209	Metal Objects (Heavenly Stems)	57
Magnesium (Element)	115	Metal, Tarot Suit	275-281
Magnesia, Greece	115	Methone (S)	191
Magpie Bridge and Lovers	277	Metis (Asteroid) First Wife of Zeus	232
Magpie represents Xǐ Shén	296, 298	Metis (J) First Wife of Zeus	170
Major Snow (Solar Term)	38	Microscopium (Constellation)	161
Makemake (Dwarf Planet)	222	Midas (Asteroid)	228
Making a Fortune (Fā Cái)	269	Military Star of North Pole (N Dip)	105
Malina Inuit Sun Goddess	197	Mimas (S) Giant	191
Manganese (Element)	119	Minerva (Athena) (Asteroid)	240
Margaret (U) Lady-in-Waiting	211	Minotaur Son of Pasiphaë	185
Maritime Silk Road (中 Centre)	284	Miranda (U) Daughter of Prospero	209
Mars	7, 16, 21	Mizar (Northern Dipper)	104
Mars (Ares) (Planet)	169	Mneme (J) (Muse)	178
Mars (Tuesday)	16	Mó Zhàng (Mara) attacks Buddha	279
Mars with Rhea Sylvia	254	Mò Zǐ, Universal Love Philosopher	268
Marsh (Earlier Heaven)	82	Mokuyōbi (Thursday)	19
Marsh (Later Heaven)	83	Molybdenum (Element)	123
Martial Arts Bā Guà	85-87	Mona (Mani)	13
Massalia (Asteroid) Marseille	233	Monday	13, 16, 18
May 4th Movement 1919 (60Y)	75	Money Tree, Immortal	273
Megaclite (J)	187	Monkey (Earthly Branch)	60
Megrez (Northern Dipper)	104	Monkey (Lunar Mansion)	30
Meitner, Lise	140	Monkey (Martial Arts)	85
Meitnerium (Element)	140	Monoceros (Constellation)	148
Melete (Muse)	175, 177	Moon (Flower Cards)	46
Mellish, John Edward	260	Moon Goddess Cháng É	21
Melpomene (Asteroid) (Muse)	230	Moon (Luna, Selene)	168
Mencius (Mèng Zǐ) Confucian Phil.	268	Moon Toad	21
Mendeleev, Dmitri Ivanovich	138	Mountain (Earlier Heaven)	81
Mendelevium (Element)	138	Mountain (Later Heaven)	84

Mountains (Heavenly Stems)	56	Niobium (Element)	123
Mouse, Four Mahjong Animals	294	Nitrogen (Element)	114
Mù Xīng (Jupiter)	7, 22	Niú Láng, Cowherd	277
Mukden (Shěnyáng) Battle (60Y)	72	Nix (Nyx) (P) Night	219
Mundilfäri (S) Norse Frost Giant	199	Nobel, Alfred	138
Musca (Constellation)	153	Nobelium (Element)	138
Muses (Jupiter)	176, 178, 181, 184, 185	Norma (Constellation)	156
Muses (Asteroids)	229, 230, 244	Northern Dipper Stars	103-105
Music Dragon, Qiú Niú	109	Nǚ Wā, Moon Goddess	21
Music, Mahjong Four Pastimes	292	Nǚ Wā, Wife of Fú Xī	7
Mysterious Woman of 9th Heaven	80	Nunki (Southern Dipper)	108
Naiad (N)	213	Nymph of the Water	4
Namaka (H) Sea Goddess, Hawaii	221	Nysa, Mount	181
Nánjīng, Great Ancient Capital	288	Oak Tree (Heavenly Stems)	55
Nánjīng, 1910 Exposition (60Y)	73	Oberon (U) King of the Fairies	210
Nánjīng, Tàipíng Rebellion (60Y)	62	Octans (Constellation)	163
Narvi (S)	200	Odin (Woden) Germanic God	14, 201
National Protection War (60Y)	74	Odysseus (Ulysses) (Asteroid)	259
Nausikaa (Asteroid)	233	Odysseus meets Calypso	192, 236
Nemausa (Asteroid) Nimes	231	Odysseus meets Nausicaa	231
Nemesis (Asteroid) Retribution	240	Odysseus with Circe	236
Neodymium (Element)	127	Oedipus, Son of Iocaste (Jocasta)	177
Neon (Element)	115	Old Dragon Head (Lǎo Lóng Tóu)	281
Neptune (Poseidon) (Planet)	213	Olympus, Mount	177
Neptune (Roman Sea God)	136	Ophelia (U) Lover of Hamlet	205
Neptunium (Element)	136	Ophiuchus (Constellation)	158
Nereid (N) Nymph of the Sea	215	Orchid, Four Mahjong Flowers	291
Neso (N) Lady of the Island	217	Orchomenus, King	178
Nessus (Asteroid) Centaur	261	Orcus (Dwarf Planet)	218
Nestor (Asteroid) Greek Warrior	257	Orion (Constellation)	146
Niad (N) Nymph of Fountains	213	Orthosie (J) (Horae)	174-177
Niǎn Rebellion (60Y)	62	Osmium (Element)	131
Nichiyōbi (Sunday)	20	Ox (Earthly Branches)	58
Nickel (Element)	119	Ox (Lunar Mansion)	27
Nightingale (Flower Cards)	40	Oxygen (Element)	114
Nimes, France (Nemausa)	231	Paaliaq (S) Inuit Shaman	195
Nine Great Emperors	104-106	Painting, Mahjong Four Pastimes	292
Nine Northern Dipper Stars	104-106	Palladium, Cassandra clings to it	124
Nine Sons of the Dragon	108-112	Palladium stolen by Diomedes	255
Níngxià, Hèlán Shān Sun God	24	Pallas (Asteroid) Pallas Athena	243
Niobe (Asteroid) loses 10 children	241	Pallene (S)	191
Niobe (Niobium) Element	123	Pampas Grass (Flower Cards)	46

Pan (S) Fertility God	189	Pisces (Zodiac)	54
Panda and Bamboo	269	Piscis Austrinus (Constellation)	162
Pandora (Asteroid) Jar of Evils	241	Plane made in China 1923 (60Y)	76
Pandora (S)	190	Platinum (Element)	132
Parisii People of Lutetia (Paris)	130	Pleiades, Daughters of Atlas	180
Parthenope (Asteroid) Siren	234	Plum Blossoms (Flower Cards)	40
Pasiphaë and Daedalus	226	Plum Blossom, Season, Flower	290, 291
Pasiphaë (J) Mother of Minotaur	184	Pluto (Hades)	136
Pasithee (J) Relaxation Goddess	183	Pluto (Hades) (Dwarf Planet)	219
Patientia (Asteroid)	247	Plutonium (Element)	136
Patroclus (Asteroid)	258	Pò Jūn (Northern Dipper)	106
Paulownia (Flower Cards)	50	Poet Ono no Michikaze (Flower C)	49
Pavo (Constellation)	159	Poland	133
Peach Blossom, Mahjong Seasons	290	Polaris (Northern Dipper)	105
Peach of Immortality	296, 300	Polis (Southern Dipper)	108
Pegasus (Constellation)	163	Polonium (Element)	133
Peony (Flower Cards)	44	Polydeuces (S) Son of Zeus	193
Perdita (U) Daughter of Leontes	208	Port Arthur (Lǚ Xùn Kǒu) (60Y)	72
Perseus (Constellation)	144	Portia (U) Wife of Bassanio	207
Petit-Prince Louis Napoleon	239	Portia (U) Wife of Brutus	207
Pheasant (Later Heaven)	83	Potassium (Element)	117
Pheasant (Lunar Mansion)	29	Praseodymium (Element)	127
Phecda (Northern Dipper)	104	Praxidike (J)	177
Pherusa (Horae)	174, 176	Priamus (Asteroid) King of Troy	256
Phobos (M)	169	Prime Minister of Qin Lǐ Sī	309
Phoebe (S) Goddess at Delphi	195	Prometheus, tortured by Eagle	128
Phoenix (Constellation)	142	Prometheus (S) tortured by Eagle	189
Phoenix (Flower Cards)	50	Promethium (Element)	128
Phoenix (Martial Arts)	86	Prospero (U) Duke of Milan	212
Phoenix & Female General Fù Hǎo	264	Protactinium (Element)	135
Pholus (Asteroid) Centaur	261	Proteus (N) Herdsman of Seals	215
Phorcys, Husband of Ceto	262	Psamathe (N) Lady of the Sand	216
Phosphorus (Element)	116	Psyche (Asteroid) marries Cupid	246
Pí Xiū (Tiān Lù) Auspicious Animal	301	Pú Láo (Bell Dragon)	111
Pichi Üñëm (Asteroid moon)	252	Pǔ Yí becomes Emperor (60Y)	73
Pictor (Constellation)	147	Puck (U) Mischievous Sprite	208
Pig (Earthly Branches)	61	Pulcova (Asteroid) Observatory	251
Pig (Later Heaven)	84	Puppis (Constellation)	148
Pig (Lunar Mansion)	28	Purple Star (Northern Dipper)	106
Pine Needles (Flower Cards)	39	Purple Forbidden City	103
Pirithous killed by Cerberus	184	Pyxis (Constellation)	150
Pisces (Constellation)	141	Qí Lín, Auspicious Animal	301

Qí Lín (Constellation)	148	Rooster (Lunar Mansion)	29
Qí Lín (Martial Arts Bā Guà)	87	Rooster (Martial Arts Bā Guà)	86
Qǐ Míng (Venus)	8	Rosalind (U) as Ganymede	207
Qīngdǎo siege by Japanese (60Y)	74	Rothschild, Baroness Bettina	250
Qiú Niú (Music Dragon)	109	Royal Uncle Cáo (Cáo Guó Jiù)	303
Quaoar (Dwarf Planet)	222	Rú Yì, Curved Wishing Bar	295, 299
Queen, Tarot	265, 269, 273, 284	Ruǎn Xiǎo Eè	264
Rabbit (Earthly Branches)	59	Rubidium (Element)	122
Rabbit (Lunar Mansion)	25	Ruò Mù Tree in the West	166
Rabbit of Jade on the Moon	21	Russian Army in Ili, Xīnjiāng (60Y)	63
Radium (Element)	134	Russia: Chinese East Railway (60Y)	69
Radon (Element)	134	Ruthenium (Element)	123
Rain (Flower Cards)	49	Rutherford, Baron Ernest	138
Rain Water (Heavenly Stems)	57	Rutherfordium (Element)	138
Rain Water (Solar Term)	33	Ryūjō Japanese ship, Taiwan (60Y)	64
Ram (Zodiac)	52	Sagitta (Constellation)	160
Rapa Nui (Easter Island)	221	Sagittarius (Constellation)	159
Rat (Earthly Branch)	58	Sagittarius (Zodiac)	54
Rat (Lunar Mansion)	27	St John's University Shanghai(60Y)	65
Red Bird of the South	11	Salacia (Dwarf Planet)	220
Red Crowned Crane, Marsh	82	Salamander of the Fire	3
Red Emperor	7	Salt Mine	115
Red Maple (Flower Cards)	48	Samarium (Element)	128
Red Star (Northern Dipper)	106	Sao (N) Lady of Rescue	216
Remus, with She-wolf	254	Saturday (Saturn)	14
Reticulum (Constellation)	145	Saturn (Cronos) (Planet)	188, 192
Rhea (S) Wife of Cronos	193	Saturn (Saturday)	14, 17
Rheinstein Castle, Germany	131	Saturn (Zhèn Xīng, Tǔ Xīng)	8, 23
Rhenium (Element)	131	Scandinavia	118
Rhine River, Germany	131	Scandium (Element)	118
Rhodium (Element)	124	Scholar, 4 Mahjong Professions	293
Rhodos, Sea Nymph	187	Scholars drink moderately	278
River Deer (Zhāng Zǐ) (Lunar M)	31	Scorpio (Zodiac)	53
Rodanthe (Nymph)	124	Scorpius (Constellation)	157
Roentgenium (Element)	140	Sculptor (Constellation)	141
Roman Days of the Week	15-17	Scutum (Constellation)	158
Romulus, with She-wolf	255	Seaborg, Glenn Teodor	138
Róng Hóng (Yung Wing) (60Y)	64	Seaborgium (Element)	139
Röntgen, Wilhelm Conrad	140	Seasons	5-6
Roof Corner Dragon, Cháo Fēng	111	Seawater (Heavenly Stems)	57
Rooftop Dragon, Chī Wěn	111	Second Revolution 1913 (60Y)	74
Rooster (Earthly Branches)	60	Sedna (Dwarf Planet)	224

-
-

Seilenos (Silenus) Satyr and Midas	226	Soil (Heavenly Stems)	56
Selene (Luna) Moon Goddess	121, 168	Sol (Sunday)	17
Selenium (Element)	121	Sol (Helios) Sun God	167
Semele (Lysithea, Thyone)	173, 178	Solar Terms	32-38
Sēn Luó Temple Battle (60Y)	70	Soldier (Chinese Chess)	311
Sēng Gé Lín Qìn (General) (60Y)	62	Solymos, Son of Chaldene	180
Serpens (Constellation)	157	Son Tay (Shān Xī Zhèn) Battle (60Y)	66
Setebos (U) Caliban's cruel God	212	Sòng Jiāng	265
Sextans (Constellation)	151	Sōng Shān (Mount Sōng)	10
Shāng Dynasty Suns	164-166	Southern Dipper Stars	107-108
Shanghai Oriental Pearl Tower	79	Spirits of the Aether (5 Elements)	4
Shanghai Securities Exch. (60Y)	75	Sponde (J) pours a drink offering	186
Shào Hào (White Emperor)	8	Spring (Seasons)	5
Shǎo Yáng (Morning, East, Spring)	79	Spring Begins (Solar Term)	33
Shǎo Yīn (Evening, West, Autumn)	79	Spring Equinox (Solar Term)	33
Shén Nóng (Yán Dì) Red Emperor	7	Spring, Seasons, Flowers	290, 291
Shěnyáng Sun Bird	165	Star(Tarot), Dǒu Mǔ North Star	279
Shì Wèi Bodyguard	308	Star Ferry Co Hong Kong (60Y)	68
Shòu Xīng, Star of Longevity	296	Star of Adversity (South Dipper)	108
Shuǐ Xīng (Mercury)	8, 22	Star of Benefit (Southern Dipper)	108
Siarnaq (S) Inuit Sea Goddess	197	Star of Birth (Southern Dipper)	108
Siberia, Allied Interve. 1918 (60Y)	74	Star of Life (Southern Dipper)	107
Siegena (Asteroid) Siegen	245	Star of Longevity (South Dipper)	108
Silenus (Seilenos) Satyr and Kalyke	182	Star of Prosperity (South Dipper)	107
Silicon (Element)	116	Statue of Liberty	120
Silk Road	272	Steam Locomotive in China (60Y)	65
Silver (Element)	124	Stephano (U) Drunken Butler	211
Sincere Star of Honesty (N Dipper)	105	Stockholm, Sweden	129
Sino-French War (60Y)	66-67	Stone Dragon Tortoise Bì Xì	111
Sinope (J) founds city on Black Sea	186	Stone Dragon Tortoise Bà Xià	111
Sisyphus carries heavy stone	131	Strabo (Asteroid) the Geographer	225
Six Southern Dipper Stars	106-108	Strength (Lion, Tarot)	277
Sixty Four Hexagrams (Yì Jīng)	87-103	String of Coins	266, 268
Sixty Year Cycle	61-76	Strontium (Element)	122
Skathi (S) Norse Winter Goddess	195	Styx (P) Goddess of River Styx	219
Sköll (S) Norse Wolf chasing Sun	197	Suān Ní (Dragon Lion)	110
Slight Heat (Solar Term)	35	Suì Xīng (Jupiter)	7, 22
Small Surplus (Solar Term)	34	Suiyōbi (Wednesday)	19
Snake (Earthly Branch)	59	Sulfur (Element)	116
Snake (Lunar Mansions)	31	Summer (Seasons)	5
Snake (Martial Arts)	86	Summer Begins (Solar Term)	34
Sodium (Element)	115	Summer Solstice (Solar Term)	35

Summer, Seasons, Flowers	290, 291	Telescopium (Constellation)	159
Sun (Flower Cards)	39	Telesto (S) "Success" (Nymph)	192
Sun and Fire (Heavenly Stems)	56	Tellurium (Element)	125
Sun Birds	23, 164-166	Tellus (Roman Earth Goddess)	125
Sūn Èr Niáng	264	Tellus (Gaea) (Planet Earth)	168
Sun God Fú Xī	23	Temperance (Tarot)	279
Sun Man at Zǐ Guǐ, Húnán	23	Temple of Heaven, Běijīng	273
Sūn Quán, King of Wú	284	Teramo, Italy (Interamnia, Aster.)	248
Sun Yat-sen kidnapped (60Y)	70	Terbium (Element)	129
Sunday (Sunne, Sol)	15	Termessos, founded by Solymos	180
Supreme Ultimate (Tài Jí)	77	Terni, Italy (Interamnia Asteroid)	248
Surtur (S) King of Fire Giants	203	Tethys (S) Wife of Oceanus	192
Suttungr (S) Norse Giant	200	Teutates (Albiorix)	195
Soot	113	Thalassa (N) Wife of Pontus	214
Swallow (Flower Cards)	49	Thallium (Element)	133
Swallow (Lunar Mansion)	27	Thallo (Horae)	174
Sword carried by Lǚ Dòng Bīn	306	The Teapot (Star Group)	106-107
Sycorax (U) Mother of Caliban	211	Thebe (J) Water Nymph	171
Sylph of the Air	3	Thelxinoe (J) (Muse)	176
Sylvia (Asteroid) Rhea Sylvia	253	Themis (Asteroid)	250
Tài Bái Jīn Xīng (Venus)	8, 22	Themisto (J) Mother of R. Danube	172
Tài Hào (Jupiter)	7	Theodora I, Byzantium	121
Tài Jí (Great Absolute)	77-97	Thisbe (Asteroid) hides from Lion	242
Tài Píng Rebellion Coin	271	Thor, Norse Thunder God	134, 197
Tài Píng Rebellion in Nánjīng (60Y)	61	Thorium (Element)	135
Tài Shān (Mount Tài)	9	Thrymr (S) King of Frost Giants	201
Tài Yáng (Sun, South, Summer)	78	Thule (Asteroid)	255
Tài Yīn (Moon, North, Winter)	79	Thulium (Element) (Thule)	130
Tān Láng (Northern Dipper)	104	Thunder (Earlier Heaven)	81
Táng Gāo Zōng, Emperor	275	Thunder (Flower Cards)	50
Tantalum (Element)	131	Thunder (Later Heaven)	84
Tantalus (Asteroid)	226	Thunor (Thor)	14
Tantalus	131	Thursday (Thunor, Thor)	14, 16, 19
Tāo Tiè (Gourmet Dragon)	110	Thyone (J) (Semele, Lysithea)	173, 178
Taoist (Daoist) Mountains	9-10	Tiān Lù (Pí Xiū) Auspicious Animal	301
Tarqeq (S) Inuit Moon God	197	Tiān Tóng Gōng (Southern Dipper)	108
Tarvos (S) Bull and 3 Cranes	198	Tián Xīng (Saturn)	8, 23
Taurus (Constellation)	145	Tiānjīn Post Office in 1878 (60Y)	65
Taurus (Zodiac)	52	Tiber River God with Aeneas	255
Taygete (J)	180	Tibetan Wheel of Life	278
Technetium (Element)	123	Tiě Guǎi Lǐ (Humility)	304
Telamon (Asteroid)	256	Tiger (Earthly Branch)	59

Tiger (Lunar Mansion)	26	Valkyrie, Battle Maiden Hilda	254
Tiger Mt Great Wall, Dāndōng	263	Vanadis (Freya) (Asteroid)	237
Timon's Banquet	207	Vanadis (Freya) Cats pull Chariot	118
Tin (Element)	125	Vanadium (Element)	118
Titan, Underworld Prisoners	118	Vanth (O) Etruscan Guide of Dead	218
Titan (S)	194	Varda (Dwarf Planet)	221
Titania (U) Queen of the Fairies	210	Varuna (Dwarf Planet)	221
Titanium (Element)	118	Vaxholm, Sweden	122, 128, 129, 130
Toghun Temür (Yuán Huì Zōng)	271	Vega (Northern Dipper)	106
Tolkien, J.R.R	221, 222	Vela (Constellation)	150
Toro (Asteroid) Bull Fight	227	Venus (Aphrodite) (Planet)	167
Tortoise and Snake	12	Venus (Tài Bái Jīn Xīng)	8, 22
Tower (Tarot) Léi Fēng Pagoda	279	Venus (Friday)	17
Treasure Bowl of Cái Shén	295, 299	Vercingetorix (Chieftain of Gaul)	120
Triangulum (Constellation)	143	Vesta (Asteroid) Hearth Goddess	231
Triangulum Australe (Constell.)	156	Victoria (Asteroid)	231
Trigrams (Bā Guà)	79- 86	Virgo (Constellation)	154
Trinculo (U) Jester of King Alonso	211	Virgo (Zodiac)	53
Triton (N) Son of Poseidon	215	Vladivostok (Hai Shen Wai)	75
Troilus (Asteroid) Son of Priam	259	Volans (Constellation)	149
Tǔ Dì Shén Earth God	168, 284	Vulpecula (Constellation)	160
Tǔ Xīng (Saturn)	8, 23	Wāng Dà Yuān sails to Africa	283
Tucana (Constellation)	163	Wáng Hǎi Lóu, Tianjin (60Y)	63
Tuesday (Tiw, Tyr)	13, 16, 18	War Dragon, Yá Zì	109
Tungsten (Element)	131	Water Dragon, Bā Xià (Gōng Fù)	112
Tutankhamun, Egyptian Pharaoh	132	Water (Earlier Heaven)	81
Twins (Gemini)	52	Water (Five Elements)	2, 4
Two Forms (Yīn and Yáng)	77	Water (Later Heaven)	84
Typhon (Asteroid) Monster	262	Water Planet (Mercury)	21
Tyrian Purple worn by Theodora I	121	Water, Tarot Suit	281- 285
Umbriel (U) Gnome of Earth	209	Wealth-attracting Child	294, 298
Undina (Asteroid) Water Spirit	252	Wednesday (Woden, Odin)	14, 16, 19
Unicorn Xiè Zhì (Lunar Mansion)	26	Wèi Bó Yáng, Alchemist	275
Urania (Asteroid) (Muse)	232	Wēi Hǎi Wèi, Battle 1895 (60Y)	69
Uranium (Element)	135	Wén Qǔ (Northern Dipper)	105
Uranus (Heaven) holds the Zodiac	135	Weywot (Q) Dancing Sky God	222
Uranus (Planet) with Wife Gaea	205	Wheel of Life (Tibet)	278
Ursa Major (Constellation)	152	White Dew (Solar Term)	36
Ursa Minor (Constellation)	155	White Emperor	8
Ursula (Asteroid) Saint, Cornwall	249	White Flower, Yī Zhàng Qīng	273
Urumqi Dragon Subduing Pagoda	79	White Pagoda, Běihǎi Park	273
Vali, changed into a Wolf	199	White Star (Northern Dipper)	104-106

White Tiger of the West	12	Xún Zǐ, Confucian Philosopher	269
Wild Boar (Flower Cards)	45	Yà Yǔ (Lunar Mansion)	28
Wild Geese (Flower Cards)	46	Yá Zì (War Dragon)	109
Willow (Flower Cards)	49	Yakub Beg, Tajik Warlord (60Y)	65
Willow Tree (Heavenly Stems)	55	Yalu River Battle (60Y)	69
Winchester (Asteroid)	246	Yán Dì Red Emperor	7
Wind (Earlier Heaven)	80	Yàn Qīng	264
Wind (Later Heaven)	84	Yán Wáng (Yama)	278
Wine Glass (Flower Cards)	47	Yān Zī Mountain in the West	165
Winter (Seasons)	6	Yáng Male Principle	78
Winter, Seasons, Flowers	290, 291	Yáng Zhōu Riot (60Y)	63
Winter Begins (Solar Term)	37	Yāo Jī Bird and Fish	267
Winter Solstice (Solar Term)	38	Yellow Dragon of the Centre	12
Wisteria (Flower Cards)	42	Yellow Emperor (Huáng Dì)	8
Wolf (Lunar Mansion)	28	Yellow Star (Northern Dipper)	105
Wolfram (Tungsten)	131	Yì Jīng 易经 64 Hexagrams	87-103
Wood (Five Elements)	1, 3	Yī Lóng River Battle (60Y)	62
Wood, Tarot Suit	263-266	Yī Zhàng Qīng, White Flower	273
Woodcutter, Four Professions	293	Yimir (S)	203
World, 7 Days 5 Elements, Zodiac	279	Yīn Female Principle	78
World Peace Coin	271	Yīn Yáng Theory	77-79
World (The World, Tarot)	280	Yíng Huò (Mars)	7, 21
Worm (Lunar Mansion)	32	Younghusband, Francis, Tibet(60Y)	70
Wú Jí Cosmic First Principle	77-79	Ytterbium (Element)	129
Wǔ Qǔ (Northern Dipper)	105	Ytterby Village, Sweden	122, 128-130
Wǔ Sōng kills a Tiger	264	Yttrium (Element)	122
Wú Sōng Road Railway (60Y)	65	Yù Gǔ Valley in the West	166
Wǔ Zé Tiān, Empress	275	Yù Tù Jade Rabbit on the Moon	21
Xenon (Element)	126	Yuán Huì Zong (Toghun Temür)	271
Xī Hé (Mother of Sun Birds)	164	Yuán Shì Kǎi Prov. President (60Y)	74
Xǐ Shén, God of Happiness	296	Yung Wing (Róng Hóng) (60Y)	64
Xī Wáng Mǔ	275	Zeus (Jupiter)	170
Xiàng Yǔ, General opposing Qín	308	Zhàn Qiáo Pier, Qīngdǎo (60Y)	68
Xiè Zhì Unicorn, Major Arcana	278	Zhāng Dào Líng, Daoist Master	276
Xiè Zhì Unicorn (Lunar Mansion)	26	Zhāng Guǒ Lǎo (Old Age)	303
Xīn Hài Revolution (60Y)	73	Zhāng Zǐ River Deer (L Mansion)	30
Xīn Yù Troop Transport (60Y)	74	Zhāo Cái Māo, Fortune Cat	302
Xú Fú sails Pacific Ocean	282	Zhāo Cái Tóng Zǐ, Child	295, 299
Xú Fū Ren, Queen of Wú	285	Zhèn Nán Pass, Battle of (60Y)	67
Xuán Dì (Mercury)	8	Zhèn Xīng (Saturn)	8, 23
Xuān Tōng Emperor Pǔ Yí (60Y)	73, 75	Zhèng Hé (1371-1433) Admiral	283
Xuán Zàng travels to India	283	Zhī Nǚ, Weaver	277

Zhílì-Fèngtiān War 1922 (60Y)　　75
Zhōng, Fā Bái 中发白　　268, 272, 283
Zhōng Lí Quán (Hàn Zhōng Lí)　　303
Zhōu Yì 周易 64 Hexagrams　　87-103
Zhū Mù Lǎng Mǎ (Mt Everest)　　113
Zhù Róng (Yán Dì, Mars)　　7
Zhū Xuān (Venus)　　8
Zhuān Xū (Mercury)　　8
Zhuāng Zǐ, Daoist Philosopher　　268
Zǐ Guǐ Sun Man, Húnán Province　　23
Zǐ Wēi (Right Assistant)　　106
Zinc (Element)　　120
Zirconium (Element)　　122
Zodiac (Western)　　51-54
Zuǒ Zōng Táng in Xīnjiāng (60Y)　　64

www.ingramcontent.com/pod-product-compliance
Lightning Source LLC
Chambersburg PA
CBHW070222190526
45169CB00001B/46